Biology in the Nineteenth Century

HISTORY OF SCIENCE

Editors

GEORGE BASALLA
University of Delaware

WILLIAM COLEMAN
Northwestern University

Biology in the Nineteenth Century:

Problems of Form, Function, and Transformation

WILLIAM COLEMAN
Northwestern University

John Wiley & Sons, Inc.
New York · London · Sydney · Toronto

Library of Congress Catalogue Card Number: 73-151725 9. 28 · 7 (

ISBN   0-471-16496-8   (cloth)
ISBN   0-471-16497-6   (paper)

Printed in the United States of America.

10   9   8   7   6   5   4   3   2   1

Series Preface

THE SCIENCES CLAIM an increasingly large share of the intellectual effort of the Western world. Whether pursued for their own sake, in conjunction with religious or philosophical ambitions, or in hopes of technological innovation and new bases for economic enterprise, the sciences have created distinctive conceptual principles, articulated standards for professional training and practice, and have brought into being characteristic social organization and institutions for research. The history of the sciences—astronomy; physics and associated mathematical methods; chemistry; geology; biology and various aspects of medicine and the study of man—consequently exhibits both great interest and exceptional complexity and presents numerous difficulties for investigation and interpretation.

For over half a century an international group of scholars has been studying the historical development of the sciences. Such studies have often called for an advanced level of scientific competence on the part of the reader. Furthermore, these scholars have tended to write for a small specialist audience within the history of science. Thus it is paradoxical that the ideas of men who are professionally committed to elucidating the conceptual development and social impact of science are not readily available to the modern educated man who is concerned about science and technology and their place in his life and culture.

The authors and editors of the *Wiley History of Science Series* are dedicated to bringing the history of science to a wider audience. The books comprising the series are written by men who are fully familiar with the scholarly literature of their subject. Their task, and it is not an easy one, is to synthesize the discoveries and conclusions of recent scholarship in history of science and present the general reader with an accurate, short narrative and analysis of the scientific activity of major periods in Western history. While each volume is complete in itself, the several volumes taken together will offer a comprehensive general view of the Western scientific tradition. Each volume, furthermore, includes an extensive critical bibliography of materials pertaining to its topics.

George Basalla
William Coleman

Contents

Biology in the Nineteenth Century

CHAPTER I

Biology

BIOLOGY WAS INTRODUCED with the nineteenth century. First came the word; a century of incessant activity was needed to create a thriving science. Biology is the study of living creatures, including the description and explanation of their structure, vital processes, and manner of production. Among natural phenomena few can be more striking than the harmonious disposition of parts and processes that make up the career of every plant and animal. From Greek antiquity the integral organism had been the principal phenomenon and fundamental problem for all who chose to study living things. This concern continued undiminished well into modern times. The plant or animal organism may be approached, however, in a variety of ways, and the definition of these special interests gave rise to distinctive doctrines, introduced novel techniques for research and exposition and, indeed, produced a specialized corps of students.

Such was the fate of biology during the nineteenth century. The term "biology" first appeared in a footnote in an obscure German medical publication of 1800. Two years later it again appeared, apparently independently, and was given ample publicity in treatises by a German naturalist (Gottfried Treviranus) and a French botanist turned zoologist (Jean Baptiste de Lamarck). The new word had gained some currency in the English language by 1820. However, biology was soon to designate one of the important and higher sciences of the Positive Philosophy of the great French social philosopher Auguste Comte, and largely through his writings of the 1830s and later propaganda by his disciples, the term won adherents and came to include under its wide shelter a host of previously disparate subjects and students.

But no term alone constitutes a science and the early definitions of

1

biology suggest limits as well as extensions to the then current studies of plants and animals. For Treviranus the

"objects of our research will be the different forms and phenomena of life, the conditions and laws under which they occur and the causes whereby they are brought into being. The science which concerns itself with these objects we shall designate Biology or the Science of Life."

Lamarck's definition read as follows:

"Biology: this is one of the three divisions of terrestrial physics; it includes all which pertains to living bodies and particularly to their organization, their developmental processes, the structural complexity resulting from prolonged action of vital movements, the tendency to create special organs and to isolate them by focusing activity in a center, and so on."

These definitions present agreement on a significant exclusion from the proper province of biology. Neither Treviranus nor Lamarck accord traditional natural history an integral place in the new science. Since the seventeenth century the description and classification of minerals, plants, and animals had prospered and progressed. A sweeping view of natural products—minerals, plants, and animals, these being contrasted with the productions of man's artifice—found a congenial home in the innumerable *Natural histories* of the eighteenth century. General descriptive activity constituted the essence of natural history and its practitioners may fairly be called Naturalists. But specialists already were active. Common usage designated students of plants Botanists and those of animals Zoologists. The attention of naturalist, botanist, and zoologist focused on external appearances, the geographical distribution of species, and the presumed relationships between different plants and animals. The principal objective of the endeavor was an ever more complete enumeration and precise and useful classification of the species of living creatures and minerals.

Those who coined the term biology were hoping to reorient the interests and investigations of all who studied life. Their foremost concern was the functional processes of the organism, those processes whose aggregate effect might well be life itself. Their concern extended physiology from medical investigations, its traditional preoccupation, to examination of the vital processes of plants and animals. William Lawrence, an English physiologist, declared that the time had come to exploit the naturalists' wealth of description, not perpetually to expand it. We must now "explore the active state of the animal [and plant] structure" and do so with the clear understanding that "observation and experiment are the only sources of our knowledge of life." With the term biology came an obvious plea to confine that science to vital functions such as respiration, generation, and sensibility. Until well into the century biology and physiology were virtually synonymous expressions.

By no means should we conclude that these claims extinguished the traditional interests and practice of the naturalist. Natural history remained a prosperous occupation throughout the nineteenth century and, toward the end of that period, was viewed by men of wider vision as an occupation rightly claiming a necessary and important share of the biologist's attention. But the ascent of plant and animal physiology was more dramatic and it offered all the appeal of a new and potentially fundamental science. Physiology itself was an ancient science and its students had often turned to animals (but, obviously, rarely to plants) for useful instruction in the workings of the human body. But physiology referred to the study of the functions of the human body and was for the most part a matter of medical concern. With few but significant exceptions little attention prior to about 1780 was accorded the vital processes of plants and animals for their sake alone. In the most tangible sense possible, physiology was wedded to medicine: most physiologists well into the nineteenth century were trained as physicians and often taught and occasionally practiced medicine as their principal livelihood. The coining of the term biology and the implications given of its global reference to all phenomena pertaining to life, whether in plant and animal or in man, is suggestive nevertheless of the subsequent development of the science. Biology during the nineteenth century, while not seriously neglecting natural history, did turn self-consciously to intensive analysis of organic functions. No less did biology gradually emancipate itself from its intellectual and institutional roots in medicine. What had been but a hopeful term in 1800 had become a vigorous and autonomous science by 1900.

Biologists and Their Institutions

Traditionally, the universities and learned academies had been the focus of scientific study in modern Europe. The quality and vigor of these institutions had varied enormously during the eighteenth century. A clear succession is evident with regard to the universities and especially their medical faculties, whose members by calling and interest evinced greatest concern for the life sciences. The Dutch university at Leiden, guided by professors whose excellent instruction was supported by distinguished research, dominated early eighteenth-century medicine. Leiden's role was later assumed by Edinburgh. Means for medical instruction and biological investigations in France were transformed by the Revolution. After 1790 Paris rivalled and then replaced Edinburgh as the Western world's center for these studies. But French hegemony lasted only into the 1840s. At that time influences from beyond the Rhine began to be felt and leadership in biology and medicine was soon to pass to the Germanies.

The German universities were perhaps the most distinctive intellectual institutions of the nineteenth century. Their impact on all realms of learning was great and, on the sciences, not least among them medicine and biology, it was overwhelming. From Germany came new ideals and a legion of inventive, superbly trained men. By the later decades of the century the German influence in biology was felt worldwide, from Russia to America, from Japan to Africa. German leadership in biology disappeared only after the double catastrophe of World War I and the Nazi purges of university and institute faculties and staff.

Medical and biological interests are not, of course, coextensive. Yet it was generally true and is easily understood that investigations now deemed predominantly biological were first pursued in a medical environment. This is evidenced by the evolving meaning of the subject and term of theoretical medicine or physiology, as noted above, and is even more apparent with regard to botany. A professorship in *materia medica* was an essential foundation in the medical faculty. It was the responsibility of the occupant of this chair to lecture on the medicinal qualities of plants, long the principal source of remedies, and often to superintend the faculty's herb garden. Over the centuries the chair of *materia medica* evolved into a post which, for all practical purposes, was one devoted to the study of plants exclusively, that is, to botany. Such was, for example, the position occupied by Carolus Linnaeus, the preeminent botanist of the modern period. The study of botany also led to investigations of lower organisms, above all microscopic organisms.

During the nineteenth century this evolution, common to most branches of learning, accelerated. The sciences were becoming specialized just as biology was defining itself as a profession. Botanist and zoologist were already specialist designations. Many more were to be added: physiologist (in the nonmedical sense), histologist, embryologist, paleontologist, evolutionary biologist, bacteriologist, and biochemist. This process has continued into the twentieth century, its pace undiminished.

Masters of these specialties and general biologists as well require, in common with the needs of all learned professions, distinctive training, sources of employment, funds, space and equipment for the pursuit of their research, facilities for the instruction of new initiates, and convenient and effective means of communication to announce discoveries and to discuss outstanding problems. Such needs placed diverse claims on society. The most obvious and perennially the least satisfied need was money. The miserable sum available for scientific work in nineteenth-century France became an open scandal and doubtlessly contributed to the precipitous decline, despite no lack of genius and effort, in quantity and overall quality of scientific work, including biology, in that nation. The leading British institutions were largely privately funded. Oxford and Cambridge encouraged mathematics, but only slowly and with extreme reluctance did they dedicate their assets to other scientific work. Effective resources for experimental biology came very late to Britain, taking its lead from the appointment in 1870 of Michel Foster to a physiological position at Trinity College, Cambridge. John Dalton (1825–1889), trained in Paris by Claude Bernard and active in New York City after 1857, was instrumental in introducing the new experimental biology to the United States. Such work, however, required considerable material assets. Experimental equipment must be purchased, laboratory space obtained, and student support provided. It was in recognition of these facts, and also from the fortunate position of possessing an ample endowment, that the creation at Johns Hopkins in 1876 of a physiological laboratory and professorship takes its importance. The new university made a major institutional commitment to biology and soon received its reward by witnessing the production of significant research and, most importantly, a remarkable generation of investigators and teachers.

Public and private money and great popular esteem had, however, long been lavished on one prominent biological institution, the museum of natural history. Museums possessing plant and animal specimens were known in antiquity and were revived by the Renaissance passion for collecting all manner of exotic objects. Botanic gardens often included collections of dried specimens; animals posed greater problems of preservation and were less favored. The early great natural history col-

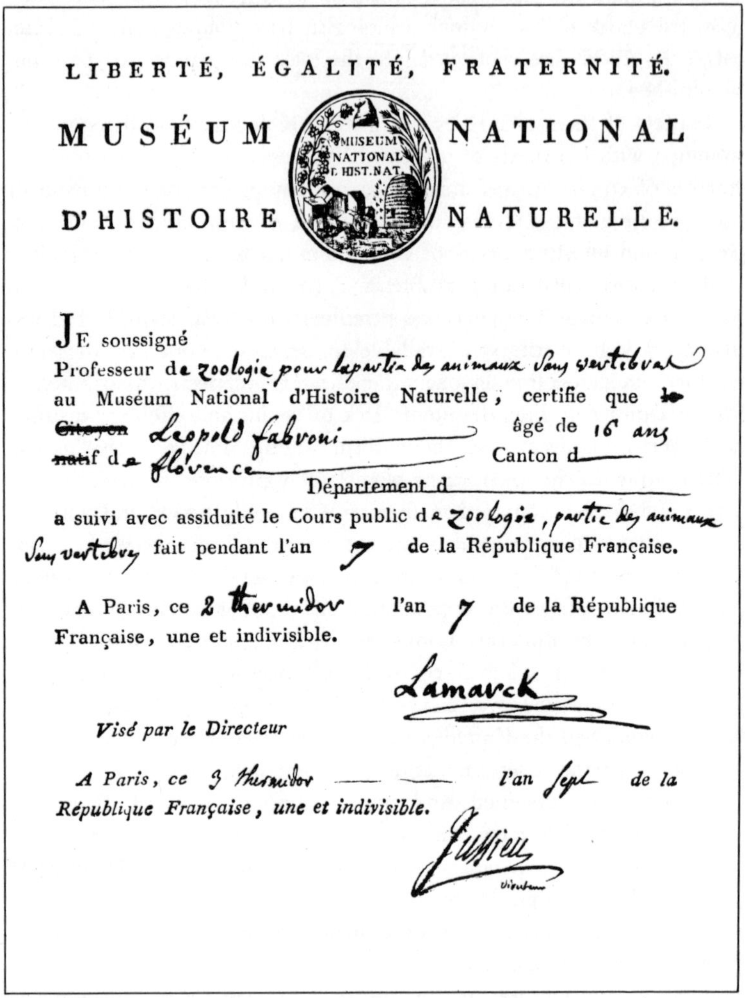

LIBERTÉ, ÉGALITÉ, FRATERNITÉ.

MUSÉUM NATIONAL

D'HISTOIRE NATURELLE.

JE soussigné
Professeur de *zoologie pour la partie des animaux sans vertébral*
au Muséum National d'Histoire Naturelle, certifie que *le*
~~Citoyen~~ *Leopold fabroni* âgé de *16 ans*
~~natif~~ d *de florence* Canton d
Département d
a suivi avec assiduité le Cours public d *a zoologie, partie des animaux*
sans vertebre, fait pendant l'an *7* de la République Française.

A Paris, ce *2 thermidor* l'an *7* de la République
Française, une et indivisible.

Lamarck

Visé par le Directeur

A Paris, ce 3 thermidor l'an *sept* de la
République Française, une et indivisible.

Jussieu

Figure 1.1 *Many natural history museums not only collected and displayed specimens but offered advanced instruction in botany and zoology. This document attests that Leopold Fabroni of Tuscany had followed the course in invertebrate zoology delivered at the Museum of Natural History in Paris by Jean Baptiste de Lamarck. It was in this famous course that Lamarck developed his evolutionary views. (American Philosophical Society.)*

lections began with national institutions for research or museum purposes. The natural history museum in Paris was founded in 1635 (as a royal botanical garden) and the British Museum was begun in 1753, its natural history collections assuming special and largely independent status in 1881. In the United States private interests led in this activity. Subscriptions by citizens of Philadelphia founded the Academy of Natural Sciences in 1812 and Louis Agassiz created Harvard's Museum of Comparative Zoology during the 1850s. A national collection became possible only in the years following the establishment in Washington of the Smithsonian Institution (1846).

All of these developments reappeared in exaggerated form in the new or revivified scientific and biological institutions supported by the various German states. Prussia was a leader in this activity. In 1809 a university, destined to become one of the greatest in the world, was created in Berlin. New foundations were also made at Breslau (1811) and Bonn (1818). The Bavarian government established a university in Munich in 1826. Its growth is indicative of the singular prosperity of the German universities. In 1826 the first appointments were made. Seventy years later Munich possessed 178 instructors of whom 98 carried professorial title. There was an enrollment of 3798 students, including 1485 in medicine and pharmacy. Four professors and 13 assistants of various rank devoted themselves exclusively to the study of living and extinct animals. A broad spectrum of special institutes had been created, funded, equipped, and staffed to pursue advanced work in zoology, physiology, paleontology, and other subjects.

The research and training institute affiliated with a university became a characteristic feature of German scientific life. It paid great scientific rewards and became the much envied model for similar foundations in other lands. Among these institutes perhaps the most conspicuous were those devoted to physiology, maintained as dependencies of the medical program of the universities. Carl Ludwig's celebrated institute in Leipzig, granted spacious and independent quarters in 1869, was designed by the physiologist himself with the special needs of his science in mind. Shaped like an E, its spine and outer wings were devoted each to a separate branch of physiology: animal experimentation, microscopical anatomy, and chemistry. The short central wing housed a lecture hall. Fully equipped laboratories were provided, a scientific library installed and assistants trained to aid both research and instruction. Indeed, it was this combined activity of original investigation and teaching that defined higher-level university work in the German institutions. Students seeking higher degrees (either M.D. or Ph.D.) participated in the professor's or institute's program of research. Their training was an inseparable part of the ongoing activity of the institute. They were research students already at work in their life's occupation. The stimulus thus provided for continued original investigation was extraordinary and there can be little wonder that after 1870, when Germany's political, economic and intellectual activity had become thoroughly disciplined, a period of work in the universities and institutes of Germany became a necessary component in the training of all who aspired to preeminence in biology.

Some of the more prominent features of the establishment of biology as a distinct member of the sciences include university positions for the teacher and, not less importantly, for his students; laboratories with adequate instruments and supplies for instruction and research; the creation of professional organizations and of specialized periodicals and other publications; the continued augmentation of museum collections and the opening up of new florae and faunae (especially, by means of seaside stations, the riches of marine life). The motives behind this activity are manifold and are yet little explored. Certainly a sentimental regard for and pleasure in nature and living things played as large a role in the development of museums as sheer acquisitiveness and national glory. Interests such as agriculture and sanitary engineering hoped to win advantages from biology and thus lent their support. Potential medical applications, as well as an integral role in the training of future physicians, encouraged growing support for physiology and other biological specialities. The vaunted ideals of the German universities, *Lernfreiheit* and *Lehrfreiheit,* the individual's freedom to learn and to teach, subject only to the control of his own good judgment, announced to the world that learning was of value in its own right, that relevance and ready applicability were not necessarily the best standards by which to judge all thought and action and that the university's endeavor, at its truest, was the highest attainment of rational men. On this elevated plane biology, too, found its proper place.

Themes and Issues in Nineteenth-Century Biology

Biological thought during the nineteenth century presents no convenient, unitary body of doctrine. While this diversity of thought constitute's biology's real vitality and interest, it precludes simplistic historical generalization. As much attention must be given to the detail and diversification of the science as is given to the elaboration of essential themes informing biology during the period. However, a suggestion of these essential themes must be presented at the outset. Historical explanation requires foremost emphasis, for this was the fundamental postulate of nineteenth-century evolutionary theorists. Its influence was felt and recognized, however, in numerous areas of biology not strictly confined to problems of the historical descent and modification of plants and animals. That influence placed its mark on the original cell theory, on the description and interpretation of the developmental changes of embryos, on evolutionary doctrines, and on theories of the nature and relationships of human society. Our point of reference need no longer be a timeless set of truths, presumably given at the Creation. Intellectual satisfaction would henceforth be found by a careful determination of antecedent conditions and ensuing consequences. The laws of nature were invariant and all natural processes capitalized on prior events. The future was being built on the acquisitions of the past. The indissoluble relation of the latter to the former constituted historical process and was deemed, in spite of earlier telling criticism by Hume and Kant, a truly causal connection. The static viewpoint, whether it demands the utter changelessness of things or the more common notion of an endless round of (undirected) cyclical historical change, seemed simply to ignore the argument and evidence produced by cosmology, geology and biology that progressive change was the most salient characteristic of natural phenomena.

"A perfect and . . . satisfactory anatomy," wrote Ignaz Döllinger (1824), a critic of the preformation doctrine and early advocate of the progressive differentiation of the embryo, "must give the time and manner of origin of all formations of the human body." The anatomist must identify the tissues of the body and notice how and when they differentiate from one another. He must also be attentive to the development of organs from these tissues. He must understand the formation of the adult body from these components and trace that body throughout its life's course. This would be a "perfect anatomy," for it related structure to all-important developmental processes. Historical or, as it was occasionally called, genetic explanation was satisfactory explanation. Thirty years later, the evolutionary philosopher *par excellence,* Herbert Spencer, complained that men, "habitually looking at things rather in their statical aspect than in their dynamical aspect, [will] never realize the fact that, by small increments of modification, any amount of modification may in time be generated." Spencer here stressed aspects of historical explanation critical to evolutionary doctrine: lack of restriction upon potential modification, the vastness of time, and the possibilities for gross change as a result of the summation of slight individual variations.

The decisive tenets of historical explanation were admirably exemplified by the rise of comparative linguistics, a subject whose conclusions and success during the nineteenth century cast valuable light on the biologist's preoccupations. Eighteenth-century students of language regarded thought as a calculus of ideas. By studying ideas one came directly to analyze the words giving them expression. A universal grammar, eternal and unvarying, was sought out and founded on the premise that mankind presented a unity in its psychological processes. The new grammar was timeless; it did not develop. To seek and, worse yet, to find a universal grammar presumably meant the denial of present and certainly of past essential linguistic diversity. Close study, however, of the structure of the Sanskrit language and its comparison with that of extinct and living west-Asian and European tongues was already suggesting to certain linguists that unmistakable degrees of relationship were attributable to familial connection, not to a universal grammar. No philologist, announced (1786) the great Orientalist, William Jones, could examine Sanskrit, Greek and Latin "without believing them to have sprung from some common source," itself perhaps now lost. Secondary linguistic peculiarities, however pronounced, could not obscure this relationship, now so firmly grounded upon common descent and the product of development. From Jones' work and the industriousness of later German linguists arose that historical school of philology whose achievements amazed the nineteenth century and to whom genetic explanation was genuine explanation. Time was indeed the measure of all things. Nature's way was neither static nor an eternal return upon past occurrences. Language, human society, and the living organisms were conceived organically. They grew. Their life-course was a record of continued eruption of novelty, divergence from expectations, and strange remnants of past circumstances. Bare reference to the timeless quality of God's well-ordered, machinelike unity seemed increasingly an inadequate explanation of this dynamic entity, Nature. Rather, one must learn how change occurs and believe that phenomena present a constant affirmation of the essentiality of change in our world. If the laws of nature were constant, the products of their play need not always be so, and the processes these laws defined were of the essence of change. Such laws and these changes presented nineteenth-century biologists with both an explanation and phenomena worthy of utmost attention. Historical explanation emphasized process, the perpetual modification of things. This emphasis expresses itself clearly in the efforts by nineteenth-century evolutionists to define the mechanism, be it environmental action or natural selection, which controlled the transformation of organisms. But this emphasis on process or change did not require the ascendancy of historical explanation alone to make it an object of intensive biological interest and investigation.

To a numerous and rapidly expanding group of biologists, historical explanation was of little interest and probably irrelevant. Their subject was physiology. Lawrence's expressed need to "explore the active state" of the organism neatly circumscribes their concern: they wished to probe ever more deeply into functional operations of the living creature. Virtually every phenomenon to greet their eye was caught up in the ceaseless flux of life, and the determination and control of that flux was the physiologist's grand objective. Physiological research advanced with startling rapidity throughout the century. By 1900 it presented abundant evidence of concrete achievement. The mystery of animal heat had long been resolved and grounds were laid for analyzing the energy relationships of life. The nature of the nervous impulse was discovered and, more significantly, conceptual and experimental means were maturing for comprehending the behavioral integrity of the organism. Chemical agents, physiologically expressed in terms of internal secretions, were found to cooperate with the nervous system in ensuring the organism's harmonious functioning. Exceptional progress had been made with regard to assaying the nature and proportions of foodstuffs required for the maintenance of life. This listing may be easily expanded and employed to demonstrate the undeniable progress, as well as the hesitations, misdirections, and errors, of physiology. To do so creates a reasonable sense of the practical accomplishments of an important science and emphasizes forcefully the specialization overtaking biology and its major subdisciplines. This record of achievement alone, however, confuses and postpones the task of isolating the common preoccupations, if such there were, of nineteenth-century physiologists.

One best seeks this ground in the realm of method and modes of explanation. Biological thought throughout the century was blessed with a bewildering variety of vitalisms and mechanisms. Behind these particular doctrines was a common goal, to announce in explicit terms what must be the ultimate being or essence of life. In defining the tasks of the new biology Treviranus remarked that the

"object of our researches is physical life. The first step towards meeting that objective must therefore be to answer the question, What is life? But just this question is the most difficult of all to answer.

One answer—pantheistic vitalism—was offered by the German naturephilosophers in the opening decades of the century. Another and fiercely opposed response was proclaimed by the radical mechanists or physiological materialists of the 1850s. Both groups ultimately borrowed their biology from their metaphysics. English physiologists in general favored a less shrill form or forms of vitalism. Materialism, mechanism, and vitalism are omnibus terms and their meanings are subject to dismaying variation; without full and explicit qualification their employment is usually pernicious.

Toward mid-century and thus contemporaneus with the most heated disputes between the rival modes of explanation emerged a self-conscious quest to make biology an experimental science. The application of experimental procedures to living organisms has a history reaching back to antiquity. The rewards of such procedures were obvious to all who had studied, for example, the reports of the Swiss physiologist Albrecht von Haller or, more strikingly, the publications of the remarkable experimental physiologists of late eighteenth-century Italy. Calmer minds among the German mechanists both advocated and fruitfully applied experimentation to biological materials. French and English physiologists would do no less.

But it was toward mid-century that this now-familiar practice came under widespread and searching inspection by the experimentalists themselves. Claude Bernard's famous *Introduction à l'étude de la médecine expérimentale* (1865)[1] was a notable systematic account. The experimentalist sought above all else rigorously to circumscribe the phenomena relevant to his interests and then to specify and exploit those terms— the variable conditions—by which the phenomena might be produced or modified. The results of suitably executed experiments could then be marshalled and general propositions ventured regarding the various bodily functions. From about 1880 experimental interests progressively gained ascendancy over biology in general. The experimentalists' work,

[1] *Introduction to the study of experimental medicine.*

as well as their campaign of publicity, spread across Europe and America and left a distinctive experimental mark upon twentieth-century biology.

In barest terms experimentation was simply a matter of manipulative procedures. It was but one method, and was called upon to become the preponderant method for biology. Most experimentalists, despite the public glory of their procedure, were not free from metaphysical commitments. In the physiological departments of German universities and institutes, where the means and impulse towards experimental work was uncommonly great, mechanism and materialism were common goods. These usually assumed the form of reductionism, whereby vital processes would be "reduced" to physics and chemistry and definite conceptual content ascribed or implied for these presumably more fundamental sciences. Bernard was philosophically less reckless, preferring to focus fullest attention on the relations between and not on the essence of biological phenomena. For his pains he found himself charged as the leader of a new vitalism.

To the more astute or temperate physiologists at century's end Treviranus' question—What is life?—had lost none of its fascination. It had, however, ceased to stand as the practical starting point for physiological research. While this large question might well continue to be the ultimate objective of physiological understanding it was displaced from the daily affairs of physiologists by more immediate concerns. The vital processes must be described and the functions, in both their independent and coordinated manifestations, analyzed. The pressing question had become how best to get along with this task. From traditional anatomy physiology borrowed and jointly exploited the venerable practice of comparison. Nature had lavishly varied her productions and keen observation, followed by careful comparison, of the different means by which certain functions were performed (for example, the support of respiration by lungs, gills, insect tracheae, or surface diffusion) could betray phyiological information of great value. Comparative physiology was viewed by some experimentalists as too passive and prey to loose analogical reasoning. Experiment proffered certainties; its scope might occasionally be narrow but its returns were dependable; its use introduced into biology the assurance that modern physics and chemistry seemed to enjoy in confronting the endless complexities of nature.

With, of course, due caution we may conclude that general physiological interests shifted significantly during the nineteenth century. Pride of place belonged always to the organism and its activities. Primacy in thought, however, shifted from efforts to define the essence of life to assiduous attention to the phenomena of life. This is best seen in the period's vital interest in questions of method and interpretation. Biology was becoming positivistic. Those who deplored the seeming intellectual aridity of this approach found their numbers diminishing and themselves often dismissed as sentimental and as too closely bound by outdated spiritual or metaphysical commitments. The sciences of life were assuredly changing their complexion.

Form, Function, and Transformation

No work of short compass can develop fairly or even cover superficially the many themes pursued by nineteenth-century biology. A severely selective plan has been followed here. What has been included should, hopefully, prove obvious but special notice must be given to the major omissions. Of these by far the most serious is the large and important realm of experimental work dealing with electrophysiology, the nature and transmission of the nerve impulse and the integrative action of the nervous system. Microbiology, the activities of the endocrine system and bodily secretions, the neurological bases of mind and the development of psychology, and doubtlessly other matters will be found absent. Certain of these omissions may prove unimportant but all were necessary. Effort has been made to discuss with some completeness the topics that are included and to ensure that, within the limits imposed, the distortions introduced will at least be recognized if not truly corrected. The limiting dates of 1800 and 1900 hold no more intrinsic meaning for the history of biology than they offer to historians in general. Overwhelming emphasis will here fall upon biological thought and practice during the nineteenth century. Wherever needed, however, and the need is common, these limits have been freely abused in order to serve the integrity of the subject and discussion.

To one group of biologists, largely comprised of anatomists, histologists, and embryologists, the appearance and constituent structures of the plant or animal body seemed all-important; they studied organic form and the means by which it was brought into being. A second body of men concentrated on the vital processes—respiration, nutrition, excretion and the like—which are diversely exhibited by all living creatures. They studied function; their self-assigned task as physiologists was to understand the inmost workings of the body. Studies of form and function were not always sharply separated and tremendous gains were to be made by pursuing a problem with the combined aid of anatomy and physiology. To a third group the problem of greatest concern was the relationship, both in the present and in past worlds, between the various kinds of plants and animals and between living beings and their changing environment. These men, later called evolutionists, studied the transformations of life over vast spans of time and, in so doing, largely recast the scientific objectives of natural history. Form, function and transformation offer, therefore, familiar and expansible vantage points from which to observe the development of the life sciences during the nineteenth century. Under these rubrics appears consideration of the structural units of life (organs, tissues, cells) and the pattern of their distribution within the organism; the developmental processes by which these units and the organisms that they constitute are formed; the recognition and statement of a satisying explanation of the changes in the form and behavior of organisms over long spans of time; the delineation of certain sciences dealing with man as an animal and as a social creature possessing an interesting and recoverable past; and the special researches and borrowings from other sciences which gave assurance that for living beings, too, energy is conserved, a demonstration that enabled physiologists confidently to designate energy as the ultimate foundation of most and probably all vital activities.

Sometime during the middle third of the eighteenth century those interested in the phenomena of life began to isolate and examine special problems for consideration and, knowingly or otherwise, to devise or articulate special techniques and viewpoints for prosecuting their examination. This process raced on undiminished throughout the nineteenth century. Its ultimate effect was to create a body of men who were recognizably biologists and whose subject, embracing a multitude of specialties, was biology. The creation of biology as a recognized discipline thus followed with only brief delay upon the determination of the legitimate subject matter of the science.

CHAPTER II

Form: Cell Theory

THE CELL THEORY offers one among several possible responses to the biologists' familiar question, What is an organism? It is a substantive response, for it presumes to describe in concrete detail the physical constitution, the structural "stuff," which constitutes the living creature. While the cell, and the larger organic structures composed of cells (for example, an organ such as the heart or spleen) or produced by cells (for example, bone or cartilage), is thus responsible for the physical makeup of the organism, the cell itself is a singularly complex structure. Minimally defined, the cell of either plant or animal will possess an exceedingly thin surrounding membrane, a nucleus containing chromosomes, and the cytoplasm, a translucent, aqueous medium filling out the cell body and containing its own specialized organelles. It is now recognized that the essential functions and hence the very life of the organism depend on the structural organization of the cell. This physiological interpretation was not an uncommon one in the nineteenth century but was advanced at that time more as a suspicion of what should be than a necessary conclusion drawn from reliable evidence.

Above the unicellular level, the living creature can be conceived as an aggregate of independent or interdependent minute anatomical elements, the cells. "Cell theory," an expression given sharp definition only towards 1840, came late to biology. It faced earlier and coherent conceptions of organic constitution and, for the most part, easily replaced these or, better, reestablished them in harmony with the new view of organic structure. The issue is largely one of levels of desirable and attainable anatomical resolution. Eighteenth-century anatomists had emphasized the structure and function of organs and systems of organs. Toward 1800 this view was challenged, principally by human anatomists who introduced the tissue doctrine. It won rapid acceptance. But one's conception of both organs and tissues was in turn to be radically transformed by the enunciation and establishment of the cell theory. After mid-century the cell had become for the great majority of biologists the essential structural reference point for the interpretation of organic form.

The bearing of cell theory reaches, however, well beyond matters of simple structural description. The ramifications of cell theory are abundant. To them we must largely assign the central position in biological doctrine which the notion of the cell and its activities has come to assume. For the cell, while always an architectural element of prime importance, is also the critical unit of organic function above the molecular level. The cell is thus the site of metabolism and energy exchange; it is the basis of nervous and secretory activity and therefore the foundation of harmonious, integrative, organic behavior; the cell, as manifested in the reproductive products, ensures, finally, the very continuity of life across the generations. The suggestion or, in the latter two instances, the discovery and demonstration of these manifold physiological capacities of the cell is virtually no other than the story of the nineteenth century's triumph and hesitation before the more general question of the functions of living creatures.

Anatomy and Organs

The anatomist could and often, of course, did pursue his art paying scant or no attention to guiding theoretical principles. Purely descriptive anatomy displayed many able practitioners. By the eighteenth century human anatomy had been brought to a remarkably high level of attainment. Human anatomists readily acknowledged the ideals of exactitude and comprehensiveness and placed particular stress on topographic studies, that is, the investigation and description of the relation of organs to surrounding organs. In the late seventeenth century interest in animal and comparative anatomy also revived. Their progress was slow but the pursuit has never since seriously slackened.

Descriptive anatomy failed nevertheless to capture the mind of all and perhaps most anatomists. It was static and recorded merely the appearance, texture, and arrangement of the parts; in extreme form it revealed nothing of the use or function(s) of those parts. To anatomists preeminent in their subject, from the Swiss physician Albrecht von Haller (1708–1777) to the Scotch comparative anatomist John Hunter

(1728–1793) and his French successor Georges Cuvier (1769–1832), anatomy was a valuable science only insofar as it recognized the necessity to pursue simultaneously both structure and function. Without a knowledge of function or the purpose for which a given organ was designed there could be no satisfactory understanding of structure itself.

Haller, Hunter, Cuvier and the many others who supported their claim were following—with vastly augmented empirical material but almost no change in the argument—the anatomical principles laid down two millenia earlier by Aristotle. "We have," Aristotle had claimed, "first to describe the common functions, common, that is, to the whole animal kingdom, or to certain large groups, or to members of a species." To the Aristotelians—the leading and growing group of biological theorists throughout the eighteenth century—anatomy, apart from its intrinsic interest, was the principal servant of physiology. The two sciences became virtually synonymous, united by a common objective: analysis of the living organism designed to explain the wholeness and harmonious interaction of the justly contrived parts of that organism.

Such was the basis of the famed *anatomia animata* of Haller and the functional anatomy of Cuvier, the latter founding a doctrine that was to guide the researches of many nineteenth-century students of organic structure. At the heart of this doctrine was the notion that one examines the parts of the body as anatomist but understands those parts as physiologist. The knowledge of structure, gained by superficial observation, dissection, and even vivisection, acquired meaning only as the "purpose" of the parts was specified. The carnivore, Cuvier argued, was perfectly constructed for his place in the economy of nature. Keen senses, great speed, and fearsome claws and teeth were nicely suited to the pursuit, capture, and consumption of animal prey. The carnivore was created to fill such a role and thereby were his construction and behavior determined.

The ultimate point of reference regarding the existence of "purpose" in nature might be the Aristotelian metaphysics and idea of the organism or, the common bond of most eighteenth-century anatomists, the Christian conception of a beneficent and all-wise creator God whose power and wisdom were readily betrayed in His works. But in either case or, as was probably true, with the support of the latter by the former, anatomists viewed form and function—the body part or organ and its necessary activities—as indissolubly joined by the idea that all existing beings were the product of intelligent and providential concern.

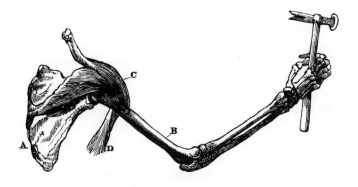

* A. The scapula, or shoulder blade; B. the humerus, or arm-bone; C. the deltoid muscle of the shoulder, arising from the shoulder-blade and clavicle, and inserted into the arm-bone; D. a muscle which draws the arm down, as in striking with a sword or hammer.

Figure 2.1 Functional anatomy (anatomia animata) *required that bodily structure and organic action receive equal consideration. The arm swinging a hammer betrays an admirable adjustment of its bones and musculature to its intended action. As a consequence of this exquisite arrangement, the organization of the arm and hand were long regarded as powerful testimony to an intelligent or creative Design manifest throughout living nature. (Charles Bell, 1833.)*

Means, the organs and thus the object of anatomy, were fitted to ends, the determination of which was the daily task and grand objective of physiology, that term being read widely to include all inquiry into organic function.

Associated with this teleological conception of the organism was the emphasis placed by anatomists on the organs and organ-systems. These parts were accessible to the unaided senses; the microscope was not yet a common or dependable aid to anatomical study. The arrangement and integrated activities of organs, made evident by topographic investigation, exhausted or at least satisfied the physiologist's search for the physical basis of bodily function. Eighteenth-century anatomists did, however, designate, but without great emphasis, more general constituents of the organism. The customary division, stated by Haller and simply repeated by Cuvier, was tripartite: nervous, muscular, and "cellular" tissue. The first two categories were most clearly defined by Haller's experimental determination of their active properties, sensibility and irritability. The third category, similar in name only to the cellular tissues described by cell theory, largely constituted the connective tissues of the body. These were described as a complex network of fibers and fibrous layers, the minute and irregular openings between these components offering a chamber- or cell-like appearance.

This view of anatomy focused attention on the particular activity or activities of the agreed-upon basic components of vital organization, organs and their assemblages. Generally speaking, concern was shown only for normal functioning. Although the abnormal or pathological was perfectly evident and of profound medical bearing, its serious and systematic study emerged only about 1760, to attain greatest prosperity after 1800. By introducing the concept of tissues, pathological anatomists were to revolutionize the concerns of traditional topographic and organ anatomy, just as the cell theory a generation later would transform pathological anatomy and all ideas of organic function.

Tissue Doctrine

Toward 1800 physicians in the Paris hospitals effected a revolution in medicine. Their essential contribution was to combine postmortem physical examination of the cadaver with a clinical description of the patient's affliction. These latter symptoms, almost exclusively the limit of eighteenth-century clinical interest, were thereby given concrete anatomical reference. Physical localization of pathological phenomena thus became the achievement and ideal of the Paris school. Philippe Pinel, spiritual father of the group, urged in the *Nosographie philosophique*[1] of 1798 that medicine should adopt, as had the other sciences, the method of philosophical "analysis," by which general or complex phenomena could be sifted, precise and well-ordered limits imposed, and district categories established. This suggested that organs that present, in health or disease, analogous phenomena must also agree in basic structure and properties. Such agreement was clearly lacking at the organ level; perhaps it might be found at a deeper level. Xavier Bichat (1771–1802) pursued this idea and, after extraordinary diligence in the dissection room, raised upon it the tissue doctrine (*Traité des membranes,* 1800; *Anatomie générale,* 1801).[2] When we study a "func-

---

[1] *Philosophical nosography.*
[2] *Treatise on membranes; General anatomy.*

tion," wrote Bichat, it is best to "examine in a general manner the complex organ which executes it." If we wish to know, however, the "properties and life" of that organ, we must "decompose it," that is, only if we "analyse [it] with rigor" can we know "its intimate structure." The study of organs was consequently but a first approximation as well as a very imperfect one to the essential truth being sought, the irreducible structural and active elements of vital organization. The latter, Bichat claimed, were the tissues.

Bichat identified and named twenty-one tissues (mucous, fibrous, cartilaginous, and so on). In addition to recording external appearance he followed the patterns of tissue distribution and used techniques of maceration and chemical reactivity in making his identifications. These and other means of analysis gave, in Bichat's skilled hands, sufficient anatomical resolution and largely compensate for his notorious neglect of the microscope. Bichat's objective in seeking and determining the various body tissues was not, of course, simple anatomical description. He recognized the manifold complexities of organ-functions and presumed that such activities must have concrete bases and that these would be seated in the tissues. The tissue doctrine was, like the Haller-Cuvier functional anatomy of organs, as much physiological as anatomical. The tissues possessed distinctive "vital properties" (sensibility and contractility, with further categories of each) and to these properties Bichat assigned both "life" itself and the diverse organ actions, the latter being a function of the particular tissues that composed the various organs. The discovery of the vital properties precluded, Bichat believed, equating life with any other natural phenomena, particularly those which were the object of physical and chemical investigation. These sciences might contribute, of course, to the progress of physiological understanding but were impotent before the grand question itself, the essential nature of vitality, of life. On such grounds numerous vitalistic doctrines were later to be erected and defended.

The tissue doctrine was announced by a physician and won greatest influence in medical anatomy. Bichat's material was the human body and students of the tissues concentrated their attention on this material. A peculiar consequence of this emphasis and its close association with medical practice and instruction was that the idea of tissues as the ultimate limit of anatomical resolution would persist, especially in France, long after the rigorous application of cell theory not only to plants and animals but also to the human frame. The cell theory later offered, in fact, some of its greatest rewards in the study of human pathology. But cell theory had to win its way against the prevailing tissue doctrine. In doing so it possessed a weapon of formidable strength and new versatility, the improved microscope.

Problems of Microscopy

The pathological anatomist, in seeking the localized disruption of tissue brought on by disease, had continued the tradition of the surgeon. His tools were the eye and the knife. By the 1850s, nevertheless, the cell theory and the microscope entered pathology. The pathologists together with all students of cell structure now happily possessed an instrument so immensely improved that serious and systematic microscopic investigation was truly feasible.

From the time of its invention in the early seventeenth century the most useful form of the instrument had remained the simple microscope (a single, roughly spherical lens in an unelaborate mounting). The image formed by compound microscopes (constructed by joining in line several separate lenses of differing shape) was vexed by spherical and chromatic aberration and other faults. The advantage of the compound microscope is that it introduces a brighter field and collects the largest feasible cone of light reflected or transmitted by the specimen, thereby permitting highest discrimination between minute objects. Without optical correction, however, these very advantages were lost in a resultant blurred field further confused by color-halos. Not until 1830 were optical theory and practice able to suggest a remedy. Then, by combining lenses of different refractive characteristics, chromatic aberration was virtually eliminated and some control imposed on spherical aberration. The earlier cell theorists used simple microscopes; by 1840, however, the improved compound instrument was available and soon became the indispensable tool of all microscopists. With it was won confirmation of the postulated cellular basis of all plants and animals, the affirmation of cellular pathology, and rectification of views on cellular generation.

Throughout the nineteenth century improvements continued to be made on the microscope. By 1880, principally through the optical investigations of Ernst Abbe and the quality manufacture of Carl Zeiss, the limits of the instrument were neared. Great magnification (about 2000 diameters) and high resolution of distinct objects (about 0.002 mm.), without which magnification becomes largely meaningless, were attainable by the oil-immersion microscope. Aberrations and distortion were reduced to a minimum, numerous microscopic techniques (fixing, sectioning, and staining) were devised, and careful instruction and practice in microscopy became the sign of modern biology. The power of the new instrument paid immediate dividends. Its capabilities, combined with appropriate preparation of materials, permitted the spectacular development in the 1880s of bacteriology and the study of subcellular structure, the one introducing a revolution in medicine and the other casting firm foundations for later explanations of inheritance.

The Cell Theory

The central propositions of the cell theory were sharply defined (1838–1839) by its articulate creators, Matthias Jacob Schleiden (1804–1881) and Theodor Schwann (1810–1882). The cell, however it is described, was the fundamental unit of organic structure and probably function. It served, consequently, as the conceptual bond that united the study of plant and animal and thereby made biology truly one science. But even the most superficial microscopic observation revealed that all cells were not alike; they were, indeed, extraordinarily varied in form and distribution. From where, then, came the conviction that the cell was the common element of life? Schleiden and Schwann grounded their belief in the idea of common cellular development, making this idea the basis of the integral cell theory. All cells were produced by an identical process. The historical argument pervading nineteenth-century biological thought operated here with full force. "The only possibility of gaining scientific insight into botany," Schleiden proclaimed (1842), "and therewith the single and indispensable methodological resource given by the nature of the object itself, is the study of organic developmental history." Schleiden and Schwann's conception of cell generation was soon shown to be erroneous, their notion of cell formation by chemical precipitation being replaced by the idea of the reproductive continuity of the living cell. Common to both views, however, and deserving of repeated emphasis, is the fact that the process of cell production was a critical point at which the meaningful assault on the problem of the general microstructure of the organism could be launched. The cell theory was not, of course, the simple creation of two inventive microscopists. It was the product of several lengthy and diverse courses of inquiry regarding organic structure and the nature of the organism. On one hand there existed a tradition of microscopical investigation coupled often with undue generalization based on frequently faulty observations; on the other hand were the highly speculative but no less suggestive conclusions of the German *Naturphilosophen* (nature-philosophers). By the 1830s these strands had merged and Schleiden and Schwann, to name only the most prominent advocates of cell theory, were subject to the influence of each. Empirical evidence and lofty doctrine were partners in cell theory. It should be recalled, moreover, that while earlier observations of cells were made with the simple microscope, after 1840 the enhanced capabilities of the compound instrument were available and freely exploited.

"Cellular" structures were first described in the seventeenth century. Microscopic examination of plants revealed discrete vesicles and also disclosed solid structures (cell walls) which, in certain plant tissues, surrounded the vesicles. Loosely speaking, "cell" merely meant a delimited "space." Structures called cells certainly existed but were deemed only one more constituent of the plant. Their manner of formation was discussed but no generally acceptable explanation was proposed. Throughout the eighteenth century these botanical observations were well known and often extended, but were the object of little serious reflection. Animal microscopy, however, faced greater obstacles. The resolution of the animal into nervous, muscular, and "cellular" (connective) tissue satisfied most observers. Animal tissues were, moreover, soft, usually low in contrast, and subject to rapid decay, all of which posed difficult problems for the microscopist still using notably imperfect instruments and techniques.

Interest in microscopic plant anatomy reawakened after 1800. Between 1800 and 1830 observers and observations multiplied and an extensive empirical foundation was gained and examined for its implications. The eminent French botanist, Charles Brisseau-Mirbel (1776–1854), for example, was at once an able microscopist and keen interpreter of his observations. He claimed that plant cells were to be found everywhere in the organism and speculated on the manner of their production. Reviving an older idea, Mirbel suggested that cells were formed *de novo* in a primitive fluid. The resulting cells and cellular tissue could be likened to the cavities formed in the foam of a fermenting liquid, the liquid by coagulation forming the continuous network of membranes so conspicuous in cellular tissue.

Mirbel's microscopic studies and speculations, including that on a formative fluid, are not unique. They are symptomatic of the study of microorganic structure during the 1820s and 1830s. Botanist and zoologist, to the regret of many, failed to share their knowledge and concerns. While plants seemed composed of cells, no one was certain what a cell might be. Its various forms were described and its diverse inclusions, most notably the cell nucleus, first remarked in 1831, were recorded. Animal tissues only slowly were subjected to dependable microscopic scrutiny. The remarkable ferment of the epoch served, however, vastly to augment the body of reliable descriptive information and to attract to consideration of the major problems of organic microstructure men of outstanding abilities.

The cell theory of Schleiden and Schwann was announced in the late 1830s, at a time when leadership in microscopy was passing from France to Germany. Young German scientists of that epoch had long been exposed to nature-philosophical doctrine and a strong case, worthy of notice, has been made that the enunciation of the cell theory was, to a significant degree, a result of such seemingly unfruitful speculations. Lorenz Oken (1779–1851), the fervent champion of nature-philosophy and one of Germany's most prolific and imaginative anatomists, most vividly illustrates this claim. Oken built upon the nature-philosophers' obsession with the fundamental problem of deriving the world's diverse productions from the inviolable unity of matter and from first principles (see Chapter III). In an aboriginal and undifferentiated mucouslike fluid (whose existence Oken postulated for the sake of argument), primitive spherical vesicles arose. Individually, each vesicle was an "infusorian" or simplest living thing; by increasing degrees of aggregation the infusoria joined together to form ever more complex organisms. The infusorian thus became the basic unit of organic structure, function, and, of course, development. "All flesh," wrote Oken (*Die Zeugung,* 1805),[3]

"may be resolved into infusorians. We can invert this statement and say that all higher animals must be formed from constituitive animalcules. These we call Primitive Animals, and note that they constitute not only the fundamental material of animals but also of plants. . . . In a larger sense they may be called the primitive matter of all which is organized."

Oken's speculations, the cause of much contemporary comment and embellishment, clearly indicate a generative element common to plants and animals, that is, a common structural and functional unit for all living things. But Oken's case was argued from metaphysics and gratuitous assumption; he spurned useful reference to concrete microanatomical examination of the organism. To those who would ground cell theory fully upon diligent empirical, microscopical investigation

[3] *On generation.*

the suppositions of Oken and his many sympathizers have always appeared bizarre and deeply harmful to sound scientific practice. To others, however, Oken's elaborations played a part in the creation of cell theory. His contributiton was not in the realm of suggestive or confirmatory evidence but was made, instead, by introducing and systematically expounding the doctrine of the composition of the living body by minute and repeated vital elements, however ill-defined or wrongly conceived these might be. Richard Owen, the distinguished English anatomist and an avowed partisan of Oken, stated as late as 1884 that "this doctrine is strikingly analogous to the generalized results of the ablest microscopical observations" on plant and animal tissues. Even in tempering Owen's generous enthusiasm, one may minimally claim that from nature-philosophy came much of the German microscopists' interest and emphatic advocacy of the central ideal that the organism may be resolved into lesser but by no means less important compositional and active units.

It was the accomplishment of Schleiden and Schwann to assess these strands of observation and thought and to put them to the test of renewed and scrupulous microscopical investigation. Their achievement was less to offer new ideas than to devise and publicize a full and coherent theory of the cells. Both were master microscopists, Schleiden (*Beiträge zur Phytogenesis,* 1838)[4] concentrating on embryonic plant tissue and Schwann (*Mikroskopische Untersuchungen über die Uebereinstimmung in der Structur und dem Wachstum der Thiere und Pflanzen,* 1839)[5] examining animal tissues (corda dorsalis and cartilage). Schleiden's investigations had convinced him that cells formed the structural basis of the plant and that these were the product of a common mode of production. These views were communicated verbally to Schwann who later recalled being "struck by the resemblance of this important body, that is, the plant cell nucleus, with a body which he [Schwann] had already often observed in animal tissues." Schwann therein suspected the foundation of a new view of organic structure, if only it could be proved that "the elementary parts of animals develop essentially in the same manner as vegetable cells." In short, if the "cause" of plant and animal generation were shown to be identical, then the products (cells) of this formative process must also be equivalent bodies.

---

[4] *Contributions to phytogenesis.*
[5] *Microscopical researches on the conformity in structure and growth between animals and plants.*

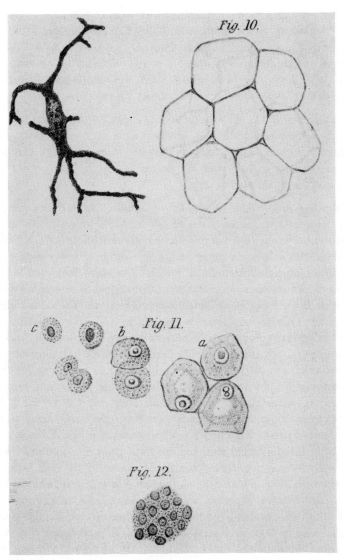

Figure 2.2   *According to the Schleiden-Schwann hypothesis of cell forma-*
tion, the cell substance is drawn from a nutritive fluid and is laid down about
particles or "nuclei" which arise in that fluid. Schwann believed this process
occurred outside of existing cells. From a cluster of nuclei can here be seen
the formation of distinctly nucleated cells and the development of large-
chambered mature cells. These observations were made on the wing feathers
of the raven. (Theodore Schwann, 1847.)

The irony of Schwann's claim was soon evident. Investigators during the 1840s and 1850s thoroughly discredited his conception of cell formation. In so doing, however, they only refounded the cell theory on more secure foundations, the idea of cell continuity via cell division. Cells, according to Schwann, who was following Schleiden's suggestion, arise in and from a formless *cytoblastema*. "There is," he stated,

"in the first instance, a *structureless* substance [cytoblastema] present, which is sometimes quite fluid, at others more or less gelatinous. This substance possesses within itself, in a greater or lesser measure according to its chemical qualities and the degree of its vitality, a capacity to occasion the production of cells."

First appeared a dark granule (the "nucleous," from which emerged the "nucleus"); successive layers of cell substance were then laid down upon it. New cells did not arise from previous whole cells. Schwann believed this process to occur principally outside of preexisting cells; Schleiden advocated intracellular new-cell formation. Both men sought, however, to make cell formation strictly analogous to inorganic crystallization. By so doing they believed they could base the science of organic form on purely physical grounds and thus be rid of the despised metaphysical excesses and vitalistic tendencies they espied in the nature-philosophical account of organic phenomena. With full allowance for the shortcomings of their proposal and sympathetic recognition of their objective it is yet a moot question whether their propositions were as radical as was proclaimed. This is particularly true with regard to the inherent and ill-explained generative quality of the cytoblastema, a conception at its roots suggestively similar to Oken's primitive mucous.

While cell formation remained the focal point of Schwann's theory of cells, he also emphasized functional aspects of the cell. Schwann asserted that the cell must be the ultimate seat of metabolic activity. Metabolic powers, however, were still very poorly understood, and Schwann's well known attempt to specify these powers was really only an exercise in hopeful expectation. He noted, nevertheless, that vital activities were limited by temperature, availability of oxygen, and presence of foreign chemical substances. The cell he deemed the focus of oxygen-carbon dioxide consumption and production. "The universality of respiration," he rather recklessly announced, "is based entirely upon this fundamental condition to the metabolic phenomena of the cells." Schwann here stated an ideal for understanding—but offered no hint of a demonstration—toward which generations of later physiologists would strive. As nineteenth-century physiology advanced from the study of the general metabolic activity of the whole organism to that of its vital elements, the cell theory was very gradually trans- formed from an essentially structural view of the organism to a prin- cipally functional interpretation of those structures.

The Schleiden-Schwann cell theory received wide publicity and largely determined the researches in microscopical anatomy until about 1860. This was a period devoted to testing the cytoblastema conception of cell formation and to extending the observational grounds for claim- ing that cells were truly a common denominator of all living creatures. Teachers of microscopic anatomy rapidly adopted the cell theory as the basis for their instruction, a fact easily verified by reference to the new and usually enormous manuals or textbooks of the subject which now became available. The authors of the most influential of these works,

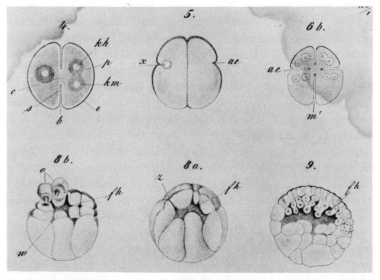

Figure 2.3   During the 1840s the distinctive stages of individual develop- ment were recognized as being composed of cells. In the next decade the process of cell formation was shown to occur uniquely by means of cell division. The observational basis for these conclusions is superbly illustrated by these drawings of stages in the development of the frog's egg, ranging from a simple two-celled body to the complex, multicellular gastrula stage. (Robert Remak, 1855.)

Jacob Henle (*Allgemeine Anatomie,* 1841)[6] and Albert von Kölliker (*Handbuch der Gewebelehre,* 1852),[7] were among the many biologists who generally supported the Schleiden-Schwann notion of cell formation. Yet, from its very announcement, the cytoblastema hypothesis had been subjected to severe criticism. This criticism came principally from microscopists who, using the new achromatic instrument, studied plant and animal structure both in the adult and developing organism. By 1860 these investigators had observed numerous instances of cell formation by the division of preexisting cells. The dramatic changes associated with embryonic development had been interpreted as being for the most part the large-scale effect of countless cell divisions that presumably began with the fertilized egg. The age-old persuasion of the continuity of life took on new and more concrete meaning, for in the cell could be seen a self-perpetuating structural and functional unit common to all living things. These discoveries, of course, contradicted and destroyed the idea of cell precipitation in a formative and nonliving cytoblastema.

Many disputes remained, even among those who accepted the new cell theory, that is, the joint conception of the ubiquity of cells in living creatures and the uninterrupted reproductive continuity, by means of cell division, of these cells. Much effort was expended, for example, in attempting abstractly to define the cell or at least determine its inalienable structures and properties. Some cytologists ruled out a membrane enveloping the cell; others strongly doubted that the cell nucleus was more than an observational artifact produced by inadequate optical apparatus or faulty preparation. Toward 1875, however, general agreement had been reached that the cell was a recognizable entity, marked off by definite spatial limits (whether or not a special bounding membrane be involved), which possessed a nucleus itself containing further specialized structures (chromosomes) and which held, most importantly, a pellucid substance (cytoplasm) of astonishing chemical and physical complexity. The period for easy generalization regarding the cell, its contents, and its vital properties was soon to give way to arduous but relentless experimental analysis of cell function and structure. This work carries one uninterruptedly into twentieth-century biology. It should be noted, however, that the "cell" as recognized in 1875 was, in its broadest specifications, very much the same "cell" so vigorously scrutinized by modern investigators.

[6] *General anatomy.*
[7] *Manual of histology.*

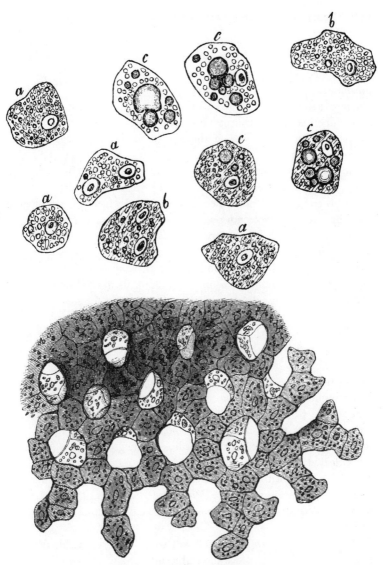

Figures 2.4 and 2.5   The textbook is often the truest key to the state of
scientific knowledge in a given period, and the conquest of biology by the
cell theory was rapid and complete. Its triumph is exemplified by these figures
of isolated human liver cells and a microsection of human liver tissue—both
taken from a major descriptive manual of the 1850s. The author continued,
however, to support the Schleiden-Schwann cytoblastema hypothesis of cell
formation. (Albert von Kölliker, 1854.)

Cell Theory and Pathological Anatomy

The triumph of cell theory was nowhere so striking as in pathology. It was largely a result of the researches and advocacy of the outspoken German medical microscopist, Rudolf Virchow, that cell theory assumed preeminent position in pathology, physiology, and general biology. Virchow (1821–1902) waged a life-long campaign against the conception—inherited from ancient Greek medicine—of "general disease," the notion that disease was an affliction of the body at large or, more precisely, of its fluids or "humors" (especially the blood). This conception Virchow replaced with his "anatomical idea." His expression was intentionally comprehensive, for by anatomical idea Virchow meant to embrace the main themes in the development of pathological anatomy since the eighteenth century, including of course the great Paris school of which Bichat is representative. Virchow's anatomical idea meant quite simply the search for the anatomical seat or seats of disease and he wished to direct pathological interest from general processes to highly localized anatomical disturbances. Students must always first ask, "Where is the disease?" The pathologist's search for the seat of disease, he declared, has now "advanced from the organs to the tissues and from the tissues to the cells." Much of Virchow's fame rests on his definition of cellular pathology. He was not original, however, in suggesting the cell as the prime point of disease. This had been frequently suggested throughout the 1840s by partisans of the cytoblastema hypothesis. But Virchow more than any other single investigator discredited this latter idea and, as part of his pathologist's mission, argued that all cells can arise only from preexisting cells (*Omnis cellula a cellula,* 1855). The cells, he wrote, are "the last constant link in the great chain of mutually subordinated formations that form tissues, organs, systems, the individual. Below them is nothing but change." Upon this proposition was founded his outstanding treatise, *Die Cellularpathologie* (1858),[8] which served to recast both the objectives and methods of pathologists' investigations.

Virchow's conception of the cell was all-inclusive and richly detailed. Following another and, since the eighteenth century, increasingly popular ancient medical proposition he argued that "disease" is really modified "life." There is no qualitative distinction between the normal and the pathological. The customary course and disposition of vital processes

---

[8] *Cellular pathology.*

and structures are, therefore, disturbed by disease but remain nonetheless the basic processes and structures upon which even disease must work its way. Virchow was thus stating a physiological definition of disease and, by introducing the cell theory, sought to delimit so far as contemporary science would allow the probable seat of disease. His definition obviously brings him to demand that, if disease is a physiological disturbance, then the cell must be the smallest and presumably irreducible organized unit of physiological activity. Cell theory had drawn together students of plants and of animals, and now Virchow sought to add pathology to their concerns. To view the diseased cell as the changed condition of the normal cell, and not as a cell of utterly different essence, forced the pathologist to attend to the disturbing conditions and to the functional response thereby engendered in the cell and cellular tissue. But such investigation was ultimately physiological. Virchow applauded this conclusion and announced that pathology, based on the cell, was not simply the application of physiology; it virtually was physiology.

In a broader and generally more profound manner Virchow reemphasized Schwann's loose claims for the functional omnipotence of the cell. Neither Schwann's suggestion nor Virchow's conviction provided the needed demonstration that the cell was the critical functional element in the organism. This demonstration remained an outstanding challenge to nineteenth-century experimental physiologists. By 1900 several lines of research, including closer examination of the respiratory processes in the organism and penetrating analysis of the structures and behavior of the nervous system, seemed to be answering that challenge. Ultimately what cell physiologists required were laboratory techniques —such as tissue or single-cell culture and microinstruments—that gave access to the intracellular components and processes themselves.

Yet nineteenth-century physiologists were far from powerless. The prolonged and arduous laboratory investigation of, for example, respiration, pursued primarily by German physiologists but given general expression in the later works of Claude Bernard (see Chapter VI), returned a new and satisfying conception of the living organism. Bernard made the cell and cellular tissues the fundamental elements in his bold delineation of the organism as a functional whole, a whole whose integral behavior depended on the dynamic interaction of the cell and the body fluids in which it bathed. If it proved exceedingly difficult to determine the particular physiological role of a given cell or cellular component, very important things could nevertheless be claimed regarding the overall role of cells and tissues in the body's functioning. The cell, demonstrably the essential structural unit of the organism, promised to become the critical functional element of the living creature. That promise bore first fruit in the study of generation and individual development.

CHAPTER III

Form: Individual Development

B ETWEEN 1840 AND 1860 biologists demonstrated that the cell was the critical organic element binding together successive generations of plants and animals. By 1876 this role had been assigned to the cell nucleus and, a decade later, competent observers had concluded that the chromosomes, distinctive formed bodies found within the nucleus, must bear ultimate responsibility for the transmission, and hence the very possibility for the continued existence, of life and living things. To the cell and, increasingly, to its constituents was ascribed the capital function of organic generation. This important conclusion was, however, less a general resolution than a more precise delimitation of age-old problems regarding generation and its concerns, notably the phenomena of heredity and variation and of individual development.

The decisive fact is, of course, that in higher forms of life "generation" is no simple event. Adult organisms (parents) do not produce directly a new adult form but only a fertilized egg. While we believe this egg and, particularly, its nucleus (both produced at fertilization by the union of the maternal and paternal reproductive products, egg and sperm, and of their nuclei) to possess remarkable molecular structure and properties, we do not suppose that the future adult form that arises from the fertilized egg exists preformed—a veritable adult in miniature —within that egg. A major advance by early nineteenth-century students of individual development (embryologists) was to destroy, so far as exacting observational evidence can do so, the then prevailing doctrine of preformation. In its place arose the idea of epigenesis. Epigenesis defines organic development as the production in a cumulative manner of increasingly complex structures from an initially more or less homogeneous material (the fertilized egg). An adult organism is thus produced in epigenetic terms by a sequence of ever-new embryonic formations, each formation building on those that went before and the whole emerging from the undifferentiated fertilized egg.

Generation, then, must include not only the reproductive act (essentially, fertilization) but will also embrace that vast complex of events that carries the developing organism from egg to adult. Epigenesis focused attention on just those events. What precisely was the course of individual development? Did organisms differ in the course of their development and, if so, what was the nature and meaning of this difference? What might be concluded regarding the relationship between these individual developmental patterns? Was there—and most nineteenth-century observers answered emphatically in the positive—an intimate connection between the development of an individual organism and the long-term historical transformations of all animals, that is, between the stages seen in individual development and those expressed in ancestral history? Finally, well after the demonstration that development truly occurs and is epigenetic, many biologists came to insist that genuine explanation of this process should be "causal" instead of merely "historical." At this point (about 1880) widespread revolt from morphology occurred and a largely new approach to problems of individual development was devised and disseminated. Nineteenth-century embryological thought thus presents its own evolution. Descriptive embryology led directly to comparative studies. From these came apparent support for the boldest of embryological generalizations, the recapitulation doctrine. Dissatisfaction with the claims and evidence of the recapitulation doctrine, which argued that the development of the individual organism will faithfully repeat the evolutionary sequence seen in ancestral history, together with new ambitions for embryological science, inspired the subsequent revolt from morphology. The study of the embryo is the study of development *par excellence*. Nowhere else in biology does the ideal of historical explanation so deeply engage itself, and embryology and its students are representative of the broadest interests and activities of nineteenth-century biology. Rare indeed was the zoological or botanical curriculum whose emphasis after mid-century did not fall toward descriptive and comparative embryology. Few biologists, certainly, would have disagreed with Carl Ernst von Baer's claim (1828) that "the history of development is the true source of light for the investigation of organized bodies."

One must note that embryology, in addition to its close conceptual bond with the general idea of development, prospered because of the very immediacy of the events it studied. A complete cycle of development was exhibited by the individual embryo and accessible therefore to all who chose to prepare themselves for its study. Nothing in the gradual evolution of plant and animal species corresponded to the beauty and seeming obviousness of the rapid embryonic transformations working themselves out before the observer's eyes. The embryo offered, furthermore, largely virgin terrain for exploitation. Its immediate fascination was matched only by the extraordinary fund of useful information it supplied to observers. The embryo and its changes were challenging problems in their own right, but the facts of individual development served other areas of biology as well. Their effect was felt most strongly in plant and animal classification, which acquired a new and seemingly fertile method when embryonic stages were added to the classifier's traditional reliance on adult characteristics (see Chapter IV).

The Bond Between Generations

The male sexual product (sperm) was discovered in the earliest years of microscopical investigation, the late seventeenth century. The female contribution (egg) was also soon known in a great variety of organisms. Throughout the eighteenth century a keen interest in the problems of generation was maintained and the catalogue of these highly diversified products expanded. But only in 1827 was the demonstration of the mammalian egg forthcoming. This demonstration (by von Baer) and the abundant earlier information provided persuasive evidence of the ubiquity of the all-important sexual products. That these products were the decisive and perhaps exclusive elements in the overall generative process was a widespread presumption. Conclusive proof for this contention was, nevertheless, lacking and numerous and often bizarre conceptions of generation repeatedly were broadcast. It was only with the introduction of cell theory and improved microscopy that concrete progress was made. Through assiduous microscopical investigation came successive establishment of the cell, nucleus, and then chromosome as the vehicle of inheritance and as the probable, if still incomprehensible, causal agent of individual development.

As early as 1843, the presence of sperm within the egg had been observed. It had already been shown that no fertilization occurs, that is, the egg does not develop, unless sperm is applied to the egg in a suitable experimental medium. Yet, until the 1870s, the preferred explanation

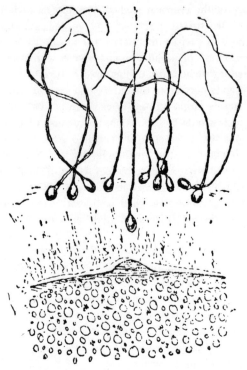

Figure 3.1   The fact that fertilization, the essential physiological event en-
suring the continuity of life in sexually reproducing organisms, required a
union of male (sperm) and female (egg) products was a common conviction
among nineteenth-century biologists. But the fertilization process itself or,
at very least, the actual penetration of the egg by a single spermatozoan, was
not easily witnessed. This figure shows a spermatozoan about to enter a ripe
egg of the starfish (Asterias glacialis) *and is regarded as the earliest repre-*
sentation of the fertilization process. (Hermann Fol, 1877.)

of the fertilization process remained the idea that the male product
contributed only molecular "excitation," a stimulus to the egg's development, and not some unspecifiable physical constituent. In 1876–1877,
Oscar Hertwig and Hermann Fol, eminent cellular microscopic anatomists (cytologists), reported, however, the undeniable presence of two
(male and female) nuclei in the fertilized egg cell. These nuclei then
merged to form the unique nucleus of the first cell of the new generation. Fol witnessed the actual penetration of the spermatozoan into the
egg. These and other discoveries discredited the notion of molecular

excitation and strongly suggested that the true meaning of fertilization would be found in the transfer of some substance, perhaps a complex molecule or group of molecules, from sperm to egg. If such were the case, then only close scrutiny of the constitution and behavior of the cell nucleus could lead to further understanding.

The nucleus had played a crucial role in Schleiden and Schwann's original cell theory. Together with its various and as yet ill-defined inclusions, the nucleus remained the object of serious microscopical attention until the late 1850s. Emphasis then shifted to another fundamental cellular component, the "living stuff" or protoplasm. But the discovery and elucidation (1873–1879) of the singular manner by which the nucleus divides (mitosis) recalled that element from neglect. What purpose, it was asked, could this seemingly inefficient and quite indirect divisional process possibly serve? Why could not simple direct division, by progressive constriction about the equator of a roughly spherical nucleus, suffice? Further studies, completed by 1883, provided an answer. The effect of indirect or mitotic nuclear division was to ensure an always equable distribution to the two daughter nuclei of the maternal and paternal contributions to the original nucleus. Nuclear division was no rude, merely quantitative halving of the massed hereditary substance; it was an exacting and necessarily intricate process which guaranteed that, disturbing factors being discounted, each cell

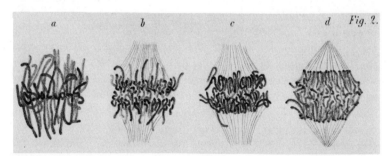

Figure 3.2 Chromosomes, darkly staining bodies detectable within the nucleus of a dividing cell, attracted great attention from microscopists after 1875. The singular behavior of chromosomes during division—most notably the duplication of the chromosome and the regular distribution of the equal products to the daughter cells—strongly suggested that these bodies were the essential elements controlling inheritance. Represented here are chromosomes in cell-nuclei of the lily embryo gathering (a,b) near the equatorial plane (c) of the dividing cell and beginning their journey (d) to the daughter nuclei and cells. (Walther Fleming, 1882.)

nucleus was identical with every other nucleus in the same organism. Here was a mechanism—one of the most magnificent of all biological processes—which provided for the precise self-duplication of cell and nucleus. Since the fertilized egg was also a cell and since it was formed by demonstrable cell products (sperm and egg), the continuity of life could be assigned first to the cell and then, more importantly, to the cell nucleus and perhaps even to its constituents.

What was the nucleus or, better, what were the essential hereditary components of the nucleus? Only in the late 1870s did answers come forth. Within the nucleus was a deeply staining substance (chromatin) which, during the phases of nuclear division, formed discrete, thread-like individually discernible bodies, chromosomes (so designated in 1888). It was the reported behavior of these distinctive and inescapable bodies that constituted the very essence of nuclear division, be it in accompaniment with normal cell division or as part of the even more complex events preceding the formation of the sexual products (meiosis). In 1884–1885 a number of leading cytologists independently concluded that the nucleus, and doubtlessly its chromosomes, was the unique and all-important physical bond between generations. From the cells and nuclei of the parent generation came specialized cells (sperm and egg) that brought forth a new and of course cellular organism, and so on *ad infinitum*.

The events just reported, and even more, those omitted from this discussion, are technical *in extremis*. Their bearing, however, is clear. A powerful case had been made by 1885 that a definite physico-chemical entity, the chromosome, was the principal and probably exclusive agent for the perpetuation of life. During the 1880s and 1890s great interest was taken in the chemical analysis of the chromosome. Estimates were made, for example, of its basic elemental constitution, and the unexpected presence of phosphorus, as well as the usual carbon, hydrogen, oxygen and nitrogen, was noted. The molecular structure of the chromosome remained, of course, an utter mystery. But how the chromosome might regulate the development of the fertilized egg and how chromosomes could be related to the suggestive but still obscure knowledge of the phenomena of heredity and variation were matters of common yet overwhelmingly speculative construction. Opportunities for rigorous experimental analysis of these problems came only in the final years of the century. The last two decades of the nineteenth century constitute, however, a period of ardent and increasingly exact study of the behavior of the integral chromosome at all moments of its existence. Of particular interest was the distribution of chromosomes during division in normal cells and in cells that were artificially disturbed by the experimentalist. The information being gathered betrayed its great importance only later and when measured against a definite conception of the distribution of hereditary qualities, this having been ascertained by breeding experiments. This comparison was possible only after the rediscovery in 1900 of Gregor Mendel's principles of heredity and variation. From about 1902 study along Mendelian lines of hereditary phenomena and the decades-long microscopical investigation of the cell, nucleus and chromosomes joined in the creation of a science of cytogenetics. Together, cytogenetics and successes in physical and physiological chemistry effected a revolution in the objectives of biological science and it is that revolution whose effects still largely define twentieth-century biology.

The Reality of Development

Nineteenth-century embryological discussion easily suggests that the perennial issue of preformation versus epigenesis had been decisively resolved in favor of the latter. While it is true that preformation won few adherents in the new century, one must not suppose that the lively debate of these alternative interpretations of development was abruptly halted. Epigenesis rightly satisfied those embryologists who claimed, following their eighteenth-century master, Caspar Friedrich Wolff, that "what [one] does not see, is not there." It demanded of them, however, doubtful propositions regarding the factors controlling individual development.

The preformationist could assume the existence and structural and functional integrity of the organism from its very beginning. The great problem facing the preformed embryo was, therefore, one of growth, of raw augmentation of the already exactly delineated embryonic creature. This was essentially a matter of nutrition and was thus coextensive with hypothetical physiological forces which allocated nutrient "particles" to appropriate parts of the developing organism. The foundations for such speculation were, of course, wholly questionable, but their conceptual roots are of great interest.

Eighteenth-century preformationists largely accepted and exploited the prescribed lawfulness of a Newtonian universe. Both the Creation and the regulative laws of the universe and of all of its lesser productions, including organisms, were ultimately the product of Divine wisdom.

Since the existence and, indeed, general and detailed form of the creature was thus referred to the First Cause, all subsequent events, that is, the actual development (really an unfolding or growth) of the embryo, were to be subordinated to one or another of the then popular systems of "mechanistic" laws. Once the hurdle of origination was surmounted, naturalists could freely assign embryonic development to secondary causation.

The task of the epigeneticist was more formidable. He was unable to postulate the aboriginal existence of any organism and was therefore compelled to account for the actual production, among the constantly changing phenomena of the present world, of such a creature. Epigenesis is change building on previous changes and in the normal course of events in the organic realm it leads always to a demonstrable and necessary end, that is, to the production of an adult organism belonging to a particular species. But the fertilized egg, these epigeneticists well knew, was truly unstructured and carried out quite independently a sequence of remarkable transformations. If, therefore, organic form is not original but is produced, what possibly can account for the regularity and directedness of such an extraordinarily complex developmental process?

To pose this question was, for most determined epigeneticists, to anticipate its answer: they postulated the existence of a special developmental force. This force acted without respite on the embryo, dictating its every transformation and ensuring that the embryo always progressed toward its goal, however indirect the actual developmental pathway might appear to be. That goal was a structurally and functionally integrated adult form. Loosely speaking, most early epigeneticists were vitalists. They required a (developmental) force active in and common to all living beings and one necessarily incommensurable with those forces of nature already subject to scrutiny and occasional control by the physical sciences, particularly mechanics. Wolff had demanded (1759) an "essential force." It was known simply "by its effect," that is, "it alone is required to explain the development of the [body's] parts." Von Baer (1792–1876), greatest of the early nineteenth-century comparative embryologists and an able proponent of epigenetic thinking, vigorously refused any mechanistic explanation of development. Instead, he claimed, it is the *"essence* (the Idea, according to the new [nature-philosophical] school) *of the developing animal form which controls the development of the germ* [fertilized egg]." Wolff's "force" and von Baer's "essence" seem to derive from different metaphysical commitments and are probably not to be equated. In common, however, they announce the early epigeneticist's need for a nonmechanistic solution to the problem of controlling the developmental process. The solution to individual organic development would have to be found in the unique properties of the organism or perhaps in its distinctive relationship to the order of nature. Von Baer's opinion was widely shared among embryologists.

A later generation fervently attacked this conclusion. Amid their attack, nevertheless, the critics had unfailingly to rely on the early epigeneticists' superb descriptions of the actual developmental process in a vast number of different organisms. Whatever was thought of the epigeneticists' varying pronouncements on developmental forces, virtually all interested in problems of individual organic development agreed that their descriptive work had established as irrefutable fact the reality of individual development. Before this evidence the traditional preformation doctrine was tried and found untenable. Von Baer's *Ueber die Entwickelungsgeschichte der Thiere* (1828–1837)[1] is the major announcement of epigenetic doctrine. Apart from its justly famous discussion of general principles for the study of organic development it represents well the interests and accomplishments of descriptive and comparative embryology, both founded on epigenetic thought.

The triumph of epigenesis begins at the University of Würzburg in the summer of 1816. There Heinrich Christian Pander (1794–1865), at the suggestion of Ignaz Döllinger and with the support of his intimate friend, von Baer, pursued in detail the development of the hen's egg from unformed primordium to hatching of the chick. The Würzburg group drew inspiration from Wolff's still largely unknown observations on this classic object of embryological research. Von Baer, however, determined to examine, compare, and report the course of development in a wide range of organisms, representatives, principally, of the various kinds of vertebrates. His investigations were always distinguished but they should only be taken as exemplary of the sound commencement and rapid advancement of the new science. By mid-century general and often highly detailed developmental histories within the major vertebrate groups (mammals, birds, reptiles, amphibians, fish) had been published. The study of the developmental patterns of the invertebrate animals had also begun. This special study, prosecuted in the rewarding environment of newly founded marine biological stations,

[1] *On the developmental history of animals.*

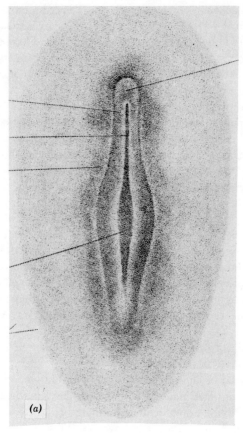

Figure 3.3 a, b, c Descriptive vertebrate embryology made extraordinary progress during the first half of the nineteenth century. A classic object for such research was the developing hen's egg. Three distinct stages (a to c) during the first 36 hours of chick development are recorded here by one of the greatest students of development and microscopical anatomy. The neural axis of body symmetry, the repeated "pairs" of future muscular tissue and the conspicuous cardial sac are clearly portrayed. (Robert Remak, 1855.)

subsequently proved one of the most vigorous of all biological disciplines.

Such impassioned descriptive work not only repeatedly confirmed the reality of individual development but led to closer attention to the texture of the developing entity, the embryo. Hereupon came the discovery of apparently exclusive embryonic tissues and also the inter-

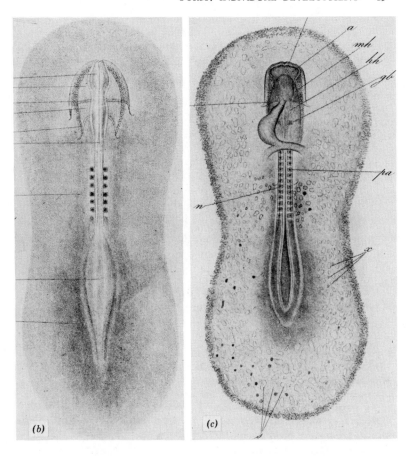

(b)   (c)

pretation of embryonic development as a strictly cellular process. Wolff, Pander, von Baer, and numerous others had remarked that embryonic development does not proceed from the egg directly to organ formation. Rather, extensive and distinctive tissue layers were formed and only by the radical transformation of these layers ("germ layers") were organs and definitive body parts later produced. Seeking the origin and tracing the fate of such embryonic tissues long remained a central preoccupation of embryology.

But the keenest students of development demanded further analysis. The germ-layers, like the organs or virtually any solid bodily constituent, must be in turn resolvable into the ultimate unit of organic structure, the cell. Between 1840 and 1855 a coherent cellular inter-

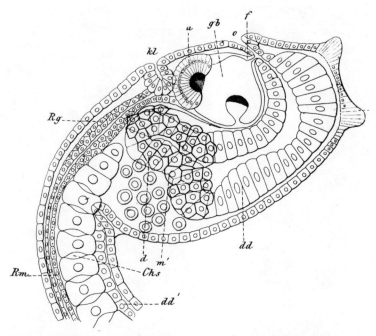

Figures 3.4 and 3.5 The emphasis of descriptive embryology shifted after 1850 toward invertebrate and principally marine animals. A vast amount of information useful for comparative studies was accumulated and the embryologist won high praise as much for the aesthetic as for the scientific interest of his work. These splendid drawings of developmental stages of an ascidian (3.4) and of Amphioxus *(3.5) are by the leading Russian practitioner of the art. (Alexander Kowalevsky, 1871, 1877.)*

pretation of vertebrate development was devised. Once again, exacting microscopical investigations by numerous observers were responsible for providing decisive evidence, this time in favor of the cellular interpretation of individual development. Robert Remak (1815–1865), the leading architect of the new conception, concluded (1855) that "at no stage in the development of the egg can a fact be presented which leads us with certainty to any other conclusion than that the independent production of nuclei or cells is really only division of preexisting nuclei or cells." Thus, in any developmental situation, the ultimate cellular components of the organism arise by "progressive division of the egg-cell into morphologically similar elements." Here, again, a major biological concern, individual development, was brought into agreement with the

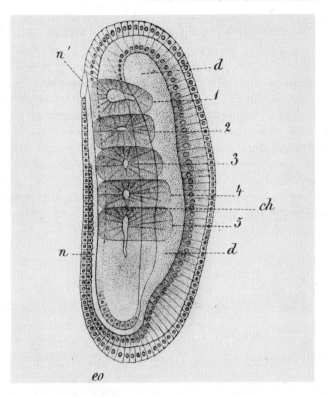

all-important cell theory. The cell here proved to be the unifying element in the historical process—individual development—best known to nineteenth-century biology.

Parallels Between Embryo and Ancestor

"Ontogeny is the brief and rapid recapitulation of phylogeny." So runs Ernst Haeckel's epigrammatic proclamation of the recapitulation theory. It argues that the course of individual development (ontogeny), especially that of higher forms of life, repeats or "recapitulates" in corresponding sequence and with suitable exactitude the progressive stages of the very evolutionary history of life on earth (phylogeny). Haeckel (1834–1919) offered this announcement in 1866 and as an integral component of a wide-ranging evolutionary synthesis. A fanatic evolutionary materialist, he was arguing "for Darwin" and found embryological themes peculiarly suited to his objective. Haeckel's efforts at interpretation and popularization of the recapitulation doctrine were remarkably successful, for this theory became a central and tremendously influential motif in post-Darwinian evolutionary thought. As a consequence, the recapitulation doctrine, at once faulty and compelling, is customarily examined in relation to the development of later evolutionary thought. Its origins, and no less the most telling criticism of its claims, go back nevertheless to earlier years of the century.

Between 1800 and mid-century the idea of organic evolutionary change remained the province of speculative natural history. Systematic and concrete supporting evidence for possible evolutionary transformations of life, above all evidence from the fossil record, was quite deficient and its accumulation was of minor interest. The prospective evolutionary record against which the various stages of a developing embryo might be compared was thus more a product of the mind than the presentation of nature. After the publication in 1859 of Darwin's *Origin of species* and the ensuing rise of interest in questions of organic transformation, the concrete trace of the evolutionary past was avidly sought and found.

Naturalists now expressed unbounded confidence that a close parallel and perhaps an identity did actually exist between the developmental series of, for example, the adult chicken from the hen's egg and that of all chicken-kind from a more primitive bird, this ancestral bird being fully traceable to the origin of the vertebrates and then on to representatives of yet more ancient creatures. Such hopes were all too sanguine. Confirmatory evidence of the parallelism was lacking where most needed; that is, no evidence showed a precise structural or sequential correspondence to exist between the embryonic and supposed ancestral stages. One may reasonably conclude that the recapitulation doctrine was built more on the expectation of this equivalence of the developmental process in both individual and race than on suggestive but treacherous evidence for that conclusion.

This interpretation shifts attention from descriptive embryology and the reconstruction of evolutionary series to the ideological roots from which the recapitulation theory grew. These roots are to be found in the metaphysics and derivative biological principles of the German *Naturphilosophen* (nature-philosophers) of the Romantic era. At the heart of the recapitulation doctrine is the idea of the identity of the forces of nature. One need not worry as to the exact delimitation of force, for "force" and "object" were inextricably joined in the nature-philosophers' conception of nature. "In so far as we regard the totality of objects," wrote Friedrich W. J. Schelling (1799), "not merely as a product, but at the same time necessarily as productive, it rises into *Nature* for us, and this *identity of the product and the productivity,* and this alone, is implied, even in the ordinary use of language, by the idea of Nature."

It was Schelling, philosophical parent of the nature-philosophers, and those of his adherents most concerned with biological matters—Oken, Döllinger, C. G. Carus and numerous others—who exceeded the cautionary limits imposed on human reason by the critical philosophy of Immanuel Kant. Schelling asked how man can "understand" nature, grasp the very essence of things, and concluded, contrary to Kant, that such knowledge was not only possible but necessary. Mind, or reason, and nature have the same source; that is, mind emerges from nature by a developmental process.

The implications of this conclusion are profound. The fundamental unity of all objects and processes was assured and the existence and ceaseless transformation of natural phenomena was to be assigned to the comprehensive term nature (again, both product and productive). To philosophize over nature was in the fullest sense of the word to create nature. For Schelling the greatest problem facing the sciences was to devise a "law" from which the phenomena of experience could be derived. Such a law was found in the principle of development, by which contrarities in nature set in shifting but opposed polar positions eternally compelled their resolution and thus dialectically moved forward the ever-developing course of nature.

The nature-philosophers' epistomological and metaphysical views are deeply interesting and inherently obscure. They provide nonetheless the essential context for understanding the remarkable qualities of these primary polar forces when applied to developmental matters of direct biological concern. If nature is fundamentally one and this unity is both produced and maintained by a no less indivisible developmental force, the tangible manifestations of such a force will obviously present striking similarities if not actual identity. In biology this meant that development, wherever it occurs, will take place along a common and reasonably well-circumscribed course. The unity of the all-controlling developmental force simply demanded this conclusion.

Already in the eighteenth-century the idea of a common developmental force had suggested uniting in one conception the then still uncertain and much disputed stages of embryonic and ancestral development. This idea was more clearly defined after 1800 and reached maturity only in the 1820s. In this latter form, the full potential of the recapitulation idea was to be exploited. The following statement (1821) by Johann Friedrich Meckel who, with the French medical anatomist, Etienne Serres, was the most explicit and thorough of the early recapitulation advocates, epitomizes the doctrine:

"The development of the individual organism obeys the same laws as the development of the whole animal series; that is to say, the higher animal, in its gradual evolution, essentially passes through the permanent organic stages which lie below it."

This fact not only assured a "close analogy" between the differences separating the various embryonic stages and those between existing animal groups. It betrayed, according to Meckel, that "tendency, inherent in organic matter, which leads it insensibly to rise to higher states of organization, passing through a series of intermediate states."

The nature-philosophers' developmental force operated under one important constraint. Given the correspondence between nature and mind the actual manifestation of nature—for the biologist, the great developmental sequence of organisms—would emerge as mind dictated. This was an idealistic philosophy and primacy of existence belonged to mind, or ideas. Each organism or, more commonly, each group of organisms was the realization of a particular idea; it was the concrete actualization, produced in time, of a timeless potential which existed eternal and unchanging in the mind of the First Cause of nature.

"Ideas" provided useful indications for animal classification. Von Baer, for example, distributed organisms in four major groups. His distribution was based on common patterns of development within each group and this pattern was in turn the result of the "essence" or idea governing the group. Embryology and animal classification of the first half of the century were penetrated by this doctrine. Ideal archetypes, reminiscent of Plato's perfect and eternal Ideas, stood above the flux of organic nature and imparted to it, for those willing to see, a coherence and logical neatness which the seemingly chaotic play of the all-generative force of nature appeared to refute. Certain later writers, principally the British anatomist, Richard Owen, and the Swiss naturalist, Louis Agassiz, attempted to reconcile the idealistic archetypal schema with the Christian conception of a Creator God who stood outside and superior to His creation. The amalgamation was difficult and not truly successful. The idea of divinely ordained and static archetypes, perhaps more than any other, cast Owen and Agassiz as the leading opponents to Darwin's notion of the self-sufficiency of the process of transformation which had produced the diversity of organic nature.

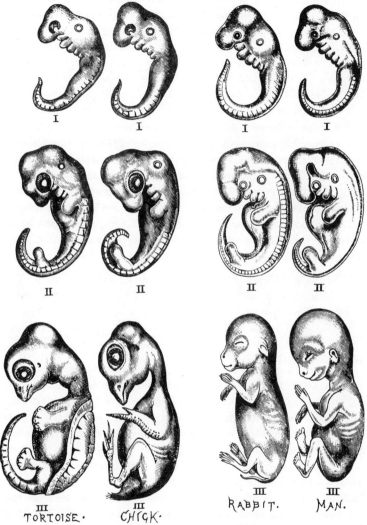

I   I   I   I

II   II   II   II

III   III   RABBIT.   MAN.
TORTOISE·   CHICK·   III   III

Figure 3.6 Nowhere did the recapitulation hypothesis appear so persuasive as in the class of the vertebrates. These striking comparisons suggest how similar are the general developmental patterns within this group and also how easy it might be, and indeed was, to claim that all vertebrates follow a single developmental pathway. The latter conclusion obscures the complex developmental specializations manifest in every species and contravenes von Baer's fundamental descriptive generalization that individual development is truly progressive differentiation. (George John Romanes, after Ernst Haeckel, 1892.)

Whether fully articulated or merely repeated *pro forma,* the re-capitulation doctrine reappeared after 1830 with familiar regularity in most treatises on anatomy, embryology, and the general principles of natural history. Agassiz (1807–1873), trained in Munich but trans-planted to Massachusetts where he helped initiate academic instruction in biology in the United States, vigorously restated the doctrine in the 1850s. He spoke of a threefold parallelism, calling attention thereby to a "genetic relation" that would explain the similarity of the sequence of life-forms in the geological record, in embryological development and in the classificatory divisions of present organisms. While sorely abusing Agassiz's conception of nature as God's handiwork, Haeckel applauded the idea of wide-ranging parallelism and, as has been noted, both ex-tended its bearing and erected recapitulation arguments into a focal dogma for evolutionary thinkers. The persistent vitality of the recapitu-lation doctrine, despite the most adverse criticism, is recorded by the following remarks—the quintessence of a century's speculation—from a standard manual of vertebrate embryology of the 1890s:

"The study of development has . . . revealed to us that each animal bears the mark of its ancestry, and is compelled to discover its parentage in its own development. . . . Evolution tells us that each animal has had a pedigree in the past. Embryology reveals to us this ancestry, because every animal in its own development repeats its history, climbs up its own genealogical tree.

Apart from its own claims, the most remarkable aspect of the re-capitulation doctrine was its survival in the face of intelligent criticism. The essential attack was delivered by von Baer in 1828. Recapitulation demanded the persistent similarity of embryos of all kinds; von Baer demonstrated that individual development is quite literally an increasing diversification. Embryos are similar only at their earliest and therefore relatively undifferentiated stage. The whole history of development leads them progressively away from one another. If this were the case, and von Baer's argument was secure, how could one rationally propose that the developing chick, for example, must exhibit embryonic stages identical with the *adult* fish, amphibian, and reptile? This had been, of course, Meckel's claim, and von Baer was cogently questioning both its logicality and its basis in evidence.

Nevertheless the subsequent history of biology proves conclusively that von Baer's widely read refutation had limited enduring effect. Historians have yet to explain why this was so. One may speculate that, until the 1850s, the idea and consequences of the nature-philosophers' dynamic view of nature—with its postulate of the unity of force and resulting processes of change—predisposed biologists to seek, and not to submit to critical analysis, the parallelism in embryonic and ancestral series. After 1859, and coincident with Haeckel's campaign, it was discovered that the recapitulation doctrine promised vast rewards in the difficult area of reconstructing the ancestral history of animals. Lacking even reasonably complete or reliable fossil series it was just too tempting to the practitioner of evolutionary reconstruction (phylogenist) to turn to the embryonic stages of a higher form, one which presumably had traversed the greatest evolutionary sequence, and find there missing ancestral stages, these being the postulated "parallels" to the present and observable embryonic stages. This argument and, above all, the promised utility of its conclusions may explain the continued prosperity of recapitulation doctrine in biology during the evolutionary period.

New Ambitions for Embryology

Recapitulation doctrine imposes an apparent unity upon nineteenth-century embryology. The claim was made that, from at least 1860, this doctrine was "the foundation on which the explanation of the facts of embryology is [to be] based." The date 1860 was merely a concession to the vagueness of the doctrine before being clarified by Haeckel's formulation and then introduced into post-Darwinian evolutionary studies. In support of the recapitulation doctrine and, occasionally, with an eye toward its refutation, exact and exhaustive investigations of the developmental stages of numerous and representative organisms were undertaken. These investigations emphasized complete and careful description, confidently assuming that positive rewards would come with incorporation of such information into a sweeping conception of ancestral-embryonic parallelism.

The limitations of this approach were increasingly evident by 1875. A decade later this low-voiced criticism was converted into a forceful program for a new approach to problems of individual development. No longer might historical explanation, drawing exclusively on descriptive and comparative embryology, suffice for understanding individual development. That understanding would henceforth come from the analysis of causal factors, meaning essentially the appropriation of physico-chemical techniques and explanatory principles heretofore neglected by morphologists. Grounds were being laid for the transformation of embryology into an experimental science.

Wilhelm His (1831–1904), a distinguished Leipzig anatomist and outspoken advocate of the new views, insisted (*Unsere Körperform und das physiologische Problem ihrer Entstehung,* 1874)[2] that the subject and methods of phylogenetic research, meaning, in embryology, the recapitulation doctrine, were "completely different" from those of the "physiological embryology of the individual." Although His fully supported traditional criticism of the recapitulation doctrine (embryonic and ancestral stages do not, as a fact, fully and exactly agree), he more willingly urged the utter irrelevance of this doctrine in gaining rigorous and useful knowledge of individual development.

For His the embryo developing before the observer's eyes was the primary fact of the science. That this embryo arose from a fertilized egg, that that egg came from the parental germ-products, that the parents themselves and the entire ancestral series were so produced were interesting facts, but facts that distracted, not served the embryologist. In the analysis of development the recreation of ancestry was, as this sketch of the great genetic continuum of life suggests, the study furthest removed from the contemporary problem of organic development. It is not surprising, therefore, that His saw in ancestral-embryonic parallelisms no contribution whatsoever to understanding the causes of individual development. "A series of forms following upon one another," he maintained, "is really, and this must be constantly repeated, no explanation." The proper form of embryological explanation would result only from the investigation and ultimate mastery of the immediate physico-chemical conditions of embryonic development. "In organic development there is," in short, "no accidental cause; every single process occupies its own peculiar place, and all together follow the order of the general periodic function of life." Embryology was thus to become a science of exactly specified time and place and had, therefore, to follow wholly as the ideals and practice of physics and chemistry, representatives of postulated genuine scientific inquiry, dictated.

His's call for "physiological morphology" was little heeded and an explicit program for the new science was to be drawn up in the 1880s and thereafter publicized by Wilhelm Roux (1850–1924). Roux used his own experiments (whose results were loudly disputed) as illustra-

[2] *Our body-form and the physiological problem of its development.*

tive material in a great campaign to redirect the science of embryology. Embryologists needed, he said, a clear-cut "question in mind" and "appropriate means of extorting an unequivocal answer to it." These investigative means might have to play on factors internal to the organism (for example, the effects of heredity and variation and the condition of the physiological fluids), or they might actually consist of the external, "environmental" circumstances in which the organism found itself. The second alternative proffered an unmistakable invitation to the experimentalist. By controlled manipulation of environmental conditions—mechanical disturbance of egg or embryo, variation of light, temperature, pressure, chemical reagents, or orientation in the gravitational or an electromagnetic field—the experimental embryologist began to produce artificially and then to codify the manifold responses of the developing organism. Such experimental work, best understood in its more mature form after 1900, began with a rush toward 1890. Embryology was entering a new and again exceptionally active phase, comparable only to the years following 1820, when descriptive work established the reality of individual development, and the post-1860 period, when embryology, exploiting the idea of ancestral-embryonic parallelism, shared the prosperity of the newly announced Darwinian conception of evolutionary change.

A faith widely shared by the new generation of embryologists was the comprehensive and, in some cases, exclusive explanatory power of their first principles. The ultimate explanation of developmental events would, hopefully, be found in the laws of physics and chemistry. These numerous embryologists, of whom His and Roux are representative, sought to capitalize on the long-established rigor and apparent certainty of the physical sciences. Roux, for example, while conceding the "complex components" necessarily present in any developmental change and hence the persistent deep obscurity of the subject, nonetheless hastened to trace each such change to a special combination of energies. He sought to found embryology on grounds as firm as those enjoyed by contemporary physics and chemistry, indeed, to view the developing organism as a truly physical system and therefore a ripe target for novel experimental analysis. But Roux and his more reflective contemporaries also recognized, given the notorious recalcitrance of the embryo to bare its secrets to any manner of analytical technique, that full satisfaction of this objective was for the future, not the present. The embryologist must, therefore, single-mindedly concentrate on his experimental work, designing it in accordance with the best procedures of the physical sciences and seeking always to win knowledge of the developmental process which could be incorporated into a broader, more physical or mechanistic scheme of things.

This objective stood at variance with that of earlier epigeneticists. It clashed sharply with von Baer's final appeal to a regulative "essence" of the developing animal and with the creation by others of definite if undemonstrable forces peculiar to the process of individual development. Roux's mechanistic persuasion also provided a weapon with which to beat down latter-day vitalists, especially Hans Driesch, whose startling publications beginning in the mid-1890s offered plausible scientific credentials to vitalists of all creeds. These issues carry embryology, and biology, into the new century and cannot be examined here. They recall, however, the shifting aspirations of nineteenth-century students of individual development. Testing the claims of epigenesis, and finding them sound, had led directly to the rise of descriptive and then comparative embryology. Virtually simultaneously came the first statement —more as a demand than a demonstration—of embryonic-ancestral parallelism. The Darwinian evolution theory revivified this doctrine, which in turn brought forth a hitherto unheard-of proliferation of embryological investigation and speculation. The essentially historical objectives of this school were, however, vehemently rejected by the new school of causal morphologists led by His and, above all, Roux. Since it failed to provide a causal explanation of development, in the form noted above, these men denied both the relevance and interest of virtually all earlier embryological investigation. This periodization of nineteenth-century attitudes toward individual organic development derives, of course, from various explicitly formulated objectives for the science. But with or without such public and demanding charges the great task of embryological description progressed dramatically during the century. Once convinced that individual development was real and not a mere appearance, the way was opened to the description of this process in virtually all accessible forms of life. It was largely the persuasion of the reality of development, brought about both by early descriptive work itself and by the widespread conviction of the dynamic potential of nature, which formed embryology into a central preoccupation and a major activity of nineteenth-century biology.

CHAPTER IV

Transformation

A LTHOUGH MATTERS PECULIAR TO NATURAL HISTORY introduced and largely defined the problem of biological species, the ramifications of this problem extended to other areas of significant and long-standing human concern. Among scientists it was the geologist who spoke most pertinently to the reflective nineteenth-century naturalist. That naturalist had come of age, moreover, in a predominantly Christian society. He understood and most commonly accepted the traditional Christian conception of a Creator God, one whose Providential concern for His creatures might be variously interpreted but never diminished or denied. Natural history derived great benefits as well as distinct liabilities from its close association, particularly in Britain, with theological interests.

Only as the century advanced did the species problem receive clear definition and capture the naturalist's foremost attention. Simultaneously with the recognition of the diversity of organic nature came an awareness of the fundamental relationship, be it of form, behavior, fecundity, or environmental situation, between numerous subunits within the great totality of endlessly varied individual organisms. Such subunits or "species" of plants and animals, however delineated, posed imperious questions. The origin and distinctness of species might be assigned to supernatural or to natural causation. Species stability over extended generations might well be doubted and conventional wisdom regarding their presumed fixity brought into question. If species indeed had a history, that history could perhaps be recovered. However unexpected or un-wanted it was, one species might actually give rise, by transformation through time, to a related but essentially distinct species and this process be repeated virtually *ad infinitum*. These and associated questions constitute the species problem. They also record the pervasive historical bias which came to predominate discussion of the species question and suggest the terms in which consideration of species transformism graduated to a fully articulated theory of the evolution of forms of life on the earth's surface.

Theological Concerns and Complications

No aspect of Darwin's evolutionary propositions evoked more concern or generated greater furor than the intimations they cast on man's biological and moral condition. Man's place in nature and, more importantly, his relation with his Creator long had been a matter for intense theological attention. As attacks on the Christian view of man and, indeed, of the Creation itself mounted ever more fiercely from the mid-seventeenth century, the theologian took good care to assess and promulgate again his doctrine. Of the many issues raised, two interrelated questions are of major concern to the historian of evolutionary doctrine.

God the Creator was indeed the great artificer so beloved of eighteenth-century deistic apologists. But He was more, a great deal more. He not only brought things into existence; His power to do so was always guided by His supreme wisdom and consummate goodness. Although these were human qualities, in their divine manifestation they became features absolutely peculiar to Himself. Man, after all, was only made in His image. He suffered, moreover, from flawing sin. God created; He was not Himself part of the creation. The universe and all creatures—nature—were thus God's handiwork and also the agent and index of His grand design. All that existed therefore served as clear testimony to God's providential concern for His creation and His fundamental nonidentity with this creation. Here was compelling argument for truly supernatural causation which had brought this world into being and presumably continued it on its course.

The theologian also emphasized the instructional and moral value of nature study. The book of nature came second only to the Bible as a guide to things divine. Natural theology, never absent from Christian thought, prospered enormously between about 1650 and 1850. It was, said Francis Bacon, "that spark of knowledge of God which may be had by the light of nature and the consideration of created things." Its objective was divine; its subject was nature. From the microscopist Jan Swammerdam (1637–1680) to the Reverend William Paley (1743–

1805), from the great naturalist John Ray (1627–1705) to the ge-
ologist-apologist William Buckland (1784–1856), generations of
naturalists described instances in nature of the divine wisdom and
power. In studying plants and animals one simultaneously increased
one's knowledge of nature and glorified nature's Creator. By 1800 the
exercise was far advanced. The relative perfection of the organism was
emphasized and imperfections largely ignored. The "purpose" of the
organism must state the conditions by which that creature could exist,
prosper, and reproduce. Such purpose, making manifest God's intention,
covered the exquisite fitness of body parts one to another and to the
functions they served, the fitness of the organism to its environment and
the providential provision of particular plants and animals for man's
every want and pleasure.

Emphasis upon purpose and perfection (as effected by the Creator)
diverted attention from, if it did not actually deny, the self-sufficiency of
man or nature to produce among organisms comparable grades of excellence. Living nature revealed how God's will had been worked, be
it at the creation or, as some modernists advised, as part of a continued
program for creation. The existence of organism and species could have
no other explanation. The idea of change in-and-by nature was, obviously, utterly alien to natural theology. Here was the crux of the
matter. However God pursued His course (Omnipotence knows no
constraint), He, and all theologians and pious naturalists who lent
Him their faith, conceded neither divorce nor unification of Creator
and creation. The former led to God's irrelevance and to materialism,
the latter to the disappearance of God's otherness and to pantheism.
Both had to be severely condemned. That plants and animals, and man,
were thus ultimately dependent upon supernatural causation, either at
the beginning or for their maintenance since that event, was the central
issue of the limitless discussion concerning evolution and theology
which broke out well before Darwin and continues to our own day.

The positive contribution of natural theology to scientific natural
history requires specification. The natural theologian, in studying design in nature, was examining and carefully describing organic adaptation, a term later to acquire technical evolutionary meaning. Our
leading principle, wrote Charles Bell (1774–1842), an outstanding
anatomist and natural theologian, is that

"there is an adaptation, an established and universal relation between
the instincts, organization, and instruments of animals on the one hand,

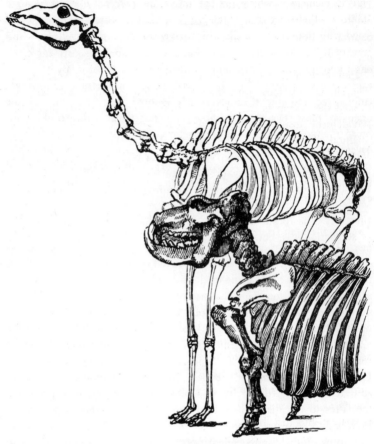

Figure 4.1   An anatomist's illustration of the adaptation of organic structure to conditions of existence, the relation being established by Creative wisdom and power: "We have here a sketch of the skeletons of the hippopotamus and of the camel, as they stood accidentally contrasted. The head of the hippopotamus is of great strength and weight, and it is appended to a short neck; in the shortness of the legs also we see the correspondence that we have had occasion to remark between the position of the head and the height of the trunk from the ground. The camel is, in every respect, a contrast. It must have rapidity and ease of motion; that is secured by the length of the extremities; and according with the extremities, are the length of the neck and the lightness of the head. Here is a skeleton, then, of an animal which is properly terrestrial, accommodated to all the other peculiarities of its organization, and adapted for a rapid and long continued course: the hippopotamus, on the other hand, seeks its safety in the water,—and its uncouth form and weight are suited to that element." (Charles Bell, 1833.)

and the element in which they are to live, the position which they are to hold, and their means of obtaining food on the other;—and this holds good with respect to animals which have existed, as well as those which now exist."

Bell here indicates the critical organism-environment relationship. The sloth, with long arms, monstrous claws and miserable gait, had evoked the "compassion" of many philosophers. Such an attitude Bell stated to be absurd. It resulted from not considering the sloth in his "natural place," that is, high among the tree branches where food, shelter and safety were his. Natural place meant adaptation, the nice adjustment of organic structure and function and the physical and bio-logical environment in which the organism and species found them-selves. Adaptation could be explained by creative foresight or by natural processes. The former was Bell's solution; the latter Darwin's. The fact of adaptation was, however, indispensable to both naturalists and offered one of the prominent facts as well as the most difficult problems in all of natural history.

The Rise of Geology

Interest in the structure of the earth's surface rapidly matured during the eighteenth century. Mineral strata were described and probable sequences of such strata established. Using this and other evidence and an overabundance of imaginative reconstruction, numerous authors turned to the creation of theories of the earth, a major aim of which was to assign a causal agency or agencies to the processes guiding the formation and distribution of geological strata. Geology soon focused upon change; it became the historical science par excellence. By 1800 two principal geological agencies had been defined and each attracted impassioned defenders.

The leading school, the neptunians, postulated the overwhelming im-portance of water, meaning usually the sea. From the primal ocean were precipitated (by means ever-obscure) all mineral strata and these were laid down in proper order. Eruptions of the sea, being sudden and violent, could also explain the many evidences—transported boulders, valley excavation, disturbed strata, and fossil remains—of major past changes on the earth's surface. The Mosaic deluge was commonly identi-fied as the most recent of such revolutions.

The vulcanist position, however, enunciated most fully by James Hut-ton about 1790, argues that the earth's central heat and internal pressures acted upon sedimentary strata and transformed and occasionally displaced them. Surface changes were strictly a result of familiar agencies (wind and water erosion, ice, and landfall) and volcanic activity. The Huttonian system took as its premise the inviolable regularity of the laws of nature; the neptunian penchant for geological catastrophes disclosed their preference for occasional or even frequent suspension or alternation of presumed lawful regularities in nature. From Hutton's insistence upon lawful regularity as the key to the earth's history arose the famous doctrine of uniformitarianism, the basis of modern geology. Only from present agencies may we legitimately reason to those of the past; between present and past (and, hopefully, future) there existed, contemporary terminology declared, an "analogy," an analogy often taken so seriously as to become an identity. "In examining things present," said Hutton, expounding his principles,

"we have data from which to reason with regard to what has been; and, from what has actually been [here an apology for the study of the geological past], we have data for concluding with regard to that which is to happen hereafter."

Supposing, therefore, that the "operations of nature are equable and steady" one can devise a standard by which to measure geological time and rates of change. All evidence pointed to the incredible slowness of such change and the consequent vastness of the time that had passed since the formation of the earth, that is, the full period during which ascertainable changes had been occurring on the earth's surface. "Time," said Hutton, "is to nature endless and as nothing" and from this opinion few if any uniformitarians cared to dissent.

It may be fairly claimed that from about 1750 onwards man's estimate of elapsed geological time proved ever-expansive, introducing therewith a revolution in his conception of the earth and its changes and of the place of plants, animals and, most significantly, man in nature. The traditional Christian position defended the literal veracity of the Mosaic account of the creation, an event certainly no more ancient than 6000 years. *Genesis* and geology stood together in mutual support well into the nineteenth century but only because compromise and increasingly imaginative readings of Scripture seemed acceptable. Nonbiblical geologists tended to concentrate exclusively on mineralogical and paleontological evidence. Buffon's estimate of 180,000 years for the age of the earth gave way to Hutton's limitless span and Charles Darwin's confessedly excessive (and unsubstantiated) 300 million years. By 1850 biologists attentive to organic transformation sensed that time was now theirs in abundance; it had ceased to be a decisive restraint upon their reconstruction of the past and the search for a mechanism of necessarily slow evolutionary change. Their satisfaction would, however, be threatened by physicists of the post-Darwinian epoch.

Strict uniformitarianism, be it that of Hutton or of its greatest and most rigorous advocate, Charles Lyell (1797–1875), was not a direct parent of biological evolution. The uniformitarian advocated an endless cycle of change, a perpetual repetition of geological destruction and reconstitution. This offered neither the development nor the emergence of genuine novelty. It was the quality (regular and without major change of degree) and mode of action (gradual and usually imperceptible) of uniform causes which so powerfully influenced the evolutionists, most notably Alfred Russell Wallace and Darwin. Evolutionary theorists, be they speculative or proudly inductive, would impose upon nature their own conception of direction or progress. The advance would therefore gain meaning and also proceed according to uniform forces.

Eighteenth-century geologists had studied not only mineral masses but fossil organic remains. Systematic description and interpretation of fossils was begun in Paris about 1800 by Georges Cuvier (1769–1832), working with extinct vertebrates, and Jean Baptiste de Lamarck (1744–1829), a student of fossil marine shells. Their work introduced a century of extraordinary paleontological activity and discovery. Paleontology presented invaluable aid to the geologist; its lesson for the historian of life was less clear and invited heated controversy. Regularities in the horizontal and vertical distribution of particular fossils in specific stratal formations quickly suggested the possibility that fossils might serve as admirable indicators of those formations, wherever geographically they might occur. On the vertical scale the fossils became the most valuable of all temporal indicators. But the time they told was strictly relative, not absolute. By themselves they could simply indicate sequence of strata, and hence the presumptive sequence of stratal formation and perhaps length of that period of formation.

To the geologist the fossil record was therefore a great asset. To the biologist that record posed more problems than it resolved. By mid-nineteenth century vast numbers of fossil forms had been described and assigned a place in what seemed a reasonable general historical succession of life on the earth. What seemed reasonable to some was, however, wholly insufficient to others. While the indubitable succession, or progression, led ever upwards towards more complex or "higher" (man)

Table 1.   Showing the Order of Superposition, or
Chronological Succession, of the Principal
European Groups of Sedimentary
and Fossiliferous Strata.

Periods and Groups.	Names of the principal Members and Mineral Nature of the Formation, in Countries where it has been most studied.	Some of the Localities where the formation occurs.

The deposits of this period are for the most part concealed under existing lakes and seas.

I. RECENT PERIOD.

A.   Consolidated sandy and gravelly beds (*a*), travertin limestones (*b*), calcareous sandstones with broken shells (*c*), coral limestone, consisting of corals, shells, &c. (*d*), compact limestone (*e*).

a. Delta of the Rhone.
b. Tivoli, and other parts of Italy.
c. Shore of island of Guadaloupe.
d. Coral reefs in Pacific, &c.
e. Bermudas.

B.   MARINE Limestone, sands, clays, sandstones, conglomerates, marls with gypsum; containing *marine* fossils (*a*).

FRESH-WATER. Sands, clays, sandstones, lignites, &c.; containing *land* and *fresh-water* fossils (*b*).

Newer Pliocene.

a. Sicily, Ischia.
b. Colle in Tuscany.

Figure 4.2 (Cont.)

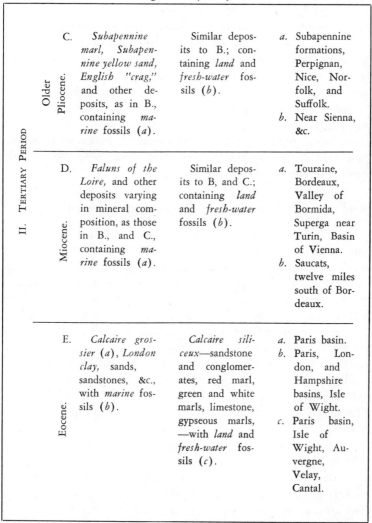

II. TERTIARY PERIOD	Older Pliocene.	C.	*Subapennine marl, Subapennine yellow sand, English "crag,"* and other deposits, as in B., containing *marine* fossils (*a*).	Similar deposits to B.; containing *land* and *fresh-water* fossils (*b*).	*a.* Subapennine formations, Perpignan, Nice, Norfolk, and Suffolk. *b.* Near Sienna, &c.
	Miocene.	D.	*Faluns of the Loire,* and other deposits varying in mineral composition, as those in B., and C., containing *marine* fossils (*a*).	Similar deposits to B, and C.; containing *land* and *fresh-water* fossils (*b*).	*a.* Touraine, Bordeaux, Valley of Bormida, Superga near Turin, Basin of Vienna. *b.* Saucats, twelve miles south of Bordeaux.
	Eocene.	E.	*Calcaire grossier* (*a*), *London clay,* sands, sandstones, &c., with *marine* fossils (*b*).	*Calcaire siliceux*—sandstone and conglomerates, red marl, green and white marls, limestone, gypseous marls, —with *land* and *fresh-water* fossils (*c*).	*a.* Paris basin. *b.* Paris, London, and Hampshire basins, Isle of Wight. *c.* Paris basin, Isle of Wight, Auvergne, Velay, Cantal.

Figure 4.2 The determination of a generalized historical sequence of geological strata was based upon the mineralogical character of the sedimentary beds and on the nature of included fossil specimens. The correlation and systematic presentation of this evidence, particularly for the Tertiary Period, was one of the principal contributions of Lyell's Principles of geology *(first edition, 1830-1833), a work that also presented in its most rigorous form the uniformitarian conception of the forces of geological change. (Charles Lyell, 1835.)*

forms, the sequence was broken by numerous gaps intolerable to the categorical evolutionist. The incompleteness of the recovered fossil record, in which a relatively full historical record for any major group was still lacking, was the very curse of the transmutationist. Life, he knew, had had a past and yet it remained beyond his grasp. Particularly was this so for crucial paleontological evidence bearing upon the transformation of a given past species into another. The gaps in the fossil record appeared to be decisive confirmation of a catastrophist view of geological change. Few paleontologists before 1859 were not ardent anti-evolutionists, and Darwin was himself compelled to devote a significant portion of the *Origin of species* to considering the imperfections of the geological record.

Evidences and Mechanisms of Evolutionary Transmutation

Technical discussion of the species problem was, despite its obvious interest to the theologian and educated man, a matter for naturalists. Their concern was with the diversity of organic nature and the merits, limitations, and rich suggestion of the systems used to give order to nature. Their conclusions bore directly, of course, on the question of whether biological species were eternal and direct from the Creator's hand or were secondarily produced, according either to God's lawful dictates or to Nature's powerful self-sufficiency. If produced, the means of production—the transmutational or evolutionary process and its mechanism(s)—must be discovered and understood. This quest was fully informed by the ideal of historical explanation and, for all matters of temporal reference, the naturalist turned confidently to the maturing science of geology.

Relationships within and between large and small groups of organisms, relationships all the more remarkable because of the appearance of irreducible diversity which first strikes the uninitiated observer—such came to be the principal fact and liveliest problem in natural history. Two areas of study were especially concerned. The first, plant and animal classification, was founded in antiquity and was ardently developed throughout the eighteenth century. The ideal of the science was natural classification, one which represented as authentically as possible the actual relationships between organisms in nature (in contrast to artificial systems which employed arbitrary and few characters and sought principally classificatory convenience). In groping toward this ideal the naturalist increasingly refined his classificatory categories. Ancient and ill-defined groups were broken up and redistributed into more natural categories. The notorious catchall unit, *Worms,* disappeared to give rise to the major subdivisions of the invertebrates. Amphibians were separated from reptiles and barnacles from molluscs. The major families of plants were delineated and further explored. This was the kind of patient work demanded by Joseph Hooker, the master of nineteenth-century British botany: "To the tyro in Natural History all similar plants may have had one parent, but all dissimilar plants must have had dissimiliar parents." Familiar experience confirmed the first position but only "years of observation" could prove the latter did not always hold true. Here was the daily round of the true naturalist.

Progress in classification exposed ever more fully the net of inter-relationships that constitutes organic nature. The call upon Providence mounted daily, for the rigorous creationist assumed that each species was the product of God's individual attention. The unveiling of the intricacies and interconnections of nature could serve either to glorify further the transcendent power and wisdom of the Creator or to press the concerned observer beyond the limits of right reason or even credulity. What a staggering amount of creative energy, it was re-marked, this world of ours seemed so necessarily to have received. It was an expression voiced by the devout, but serviceable as well to the sceptic.

Still more remarkable evidence bearing on the pattern of nature was contributed by the study of plant and animal distribution. Diverse interests had carried European explorers into all corners of the globe, and among the riches they returned home not least were abundant plant and animal specimens. These objects were, moreover, studied in the field and detailed accounts prepared. Explicit biogeographical study begins only about 1750. Significant regularities in the distribution of organisms were soon discerned: the relatedness of florae and faunae on islands in archipelagoes, the gradual replacement of predominant life forms as one traverses a great continental land mass, the marked peculiarities of plant and animal populations cut off from other like forms by prominent surface features, the distinctive mountainous life zones which seemed a function of elevation, and numerous other instances.

These patterns demanded explantion. Lyell saw clearly that plant and animal distribution was intimately related to surface conditions. Geology taught that the latter were slowly but constantly changing; biogeographical understanding must therefore also cast its answers in terms of change. Species would spread more or less widely from centers of creation, ever subject to ambient conditions.

"The possibility of the existence of a certain species in a given locality," Lyell wrote (1832), "or of its thriving more or less therein, is determined not merely by temperature, humidity, soil, elevation, or other circumstances of the like kind, but also by the existence or nonexistence, the abundance or scarcity of a particular assemblage of other plants and animals in the same region."

The distribution of species and indeed of all biological groups is thus subject to manifold and extreme vicissitudes. Lyell's argument showed that such distribution is largely if not wholly the consequence of historical events. The existence of geographical patterns and the peculiar facts associated with certain of these patterns became the cornerstone for the evolutionary theories of both Wallace and Darwin. Lyell himself was a temperate creationist who denied the mutability of species and thus assigned the ultimate plan of nature to its Creator. His view was the prevailing but not exclusive view among naturalists until 1859. No evidence, however, so fully confirms the fact of evolutionary change as does that from biogeography, and no evidence so amply derives its modern explanation from that same doctrine.

These accumulating evidences of evolutionary descent acquired full meaning only after publication of the *Origin of species* (1859). Early nineteenth-century discussion of the transmutation doctrine was essentially speculative. The marshalling of evidence for descent with modification was a secondary matter to authors preoccupied by the ostensible causal mechanisms of species transmutation. At the basis of such speculation, to be illustrated here by the views of the distinguished French naturalist, Lamarck, and those in the doctrinaire *Vestiges of the natural history of creation* (by Robert Chambers but published anonymously in 1844) stood total allegiance to historical explanation and the conviction that such explanation was no mere mental convention but a veritable representation of natural processes. "There exists throughout the universe," Lamarck asserted, "an astonishing activity which no cause can diminish, and everything which exists seems constantly subject to necessary change." Thus, in nature there really exist no such things as families, orders, genera, or constant species of plants and animals. All existence is nothing other than eternal becoming, a vast flux of evanescent qualities and unstable entities. Even classificatory categories dissolve before the insistent dynamism of nature.

Lamarck's remarkable evolutionary scheme then postulated the mechanism by which such change permeates the organic world. Developing a characteristic form of eighteenth-century materialism, Lamarck united life and matter by means of the action of the subtle fluids (heat and light) and thus endowed matter with "activity." Activity rendered the organism responsive to the environment. As the environment changed, so changed the organism, the whole process receiving elaborate physiological and psychological parameters. "Favorable circumstances," therefore, and "time," acting on the malleable organism, were the essential transforming forces; there were no limits to time and circumstances were ever-varying. Lamarck was a categorical geological uniformitarian.

The notion of all-pervasive change explained for Lamarck the many evident departures from a unilinear progressive scale of living creatures. This scale (*scala naturae* or great chain of being) derived from the Platonic notion of cosmic plenitude and the Aristotelian doctrine of hierarchies. Harmoniously integrated into Christian thought, it represented man as the middle term in a great series of existences sweeping from inorganic nature through lower to higher plants and to animals and on beyond man to the heavenly realm of angels and God the Creator. While paying due reverence to its uppermost terms naturalists were most fascinated by the seeming harmonies and symmetry of the scale in its application to the mundane representatives of the organic kingdoms. The scale of nature dominated eighteenth-century biological thought. It could be viewed ordinally (that is, as presently complete and static), or temporally (that is, as the Creator's unfolding plan for His creation). The latter became increasingly popular and presents obvious relationship with the ideal of historical explanation and the bare notion of organic transformation.

Both the temporal and ordinal scales, in their rigorous form, postulate the unmodified and uninterrupted ascent of life. The hard facts of natural history, however, confound this idea. Cuvier's system of classification shattered all aspirations for an ordinal scale; Lamarck's conception of environmental influences was designed to explain recognized departures from the ideal temporal scale. Such departures existed and could no longer be denied. Lamarck hoped to preserve the idea of a generalized progression of life forms and simultaneously account for the specialization and patent adaptation to circumstances of the great array of plant and animal species. While contemporary judgments on his transmutational mechanism were almost uniformly negative, the exposition of that mechanism points out once again the critical role that organism-environment relationships have played in the development of evolutionary doctrine. It appears that no serious naturalist has ever rested content with unexplained diversity in nature.

This doctrine Lamarck first enunciated in 1800, expanded in 1809 (*Philosophie zoologique*)[1] and brought to fullest development in 1815. Lamarck was a superb and an honored botanist and zoologist. His speculative writings, however, met either unseeing eyes or ill-tempered critics. Concrete evidence favoring his scheme was at best minimal. His premises, furthermore, demanded the faith of a metaphysician, not the presumed observational impartiality of the scientist. Worst of all, Lamarck perpetuated dogma too closely associated with Enlightenment radicalism into a period of massive Europe-wide conservative political and intellectual reaction.

But however muted or even disguised, his was the voice of the evolutionary partisan in the pre-Darwinian period. "Little by little," he had said, "nature has succeeded in forming the animals such as we see them before us." Organic diversity was assigned indubitable natural causation. The critics knew their enemy well. Lyell saw fit to devote a full volume to refuting Lamarck's doctrine; Buckland insisted that the paleontological record simply contradicted Lamarckian presumption; Edward Hitchcock, an eminent American geologist and divine, contended *a propos* of Lamarck that the "occurrence of events according to law does not remove the necessity of a divine contriving, superintending and sustaining Power."

The clamor over the transmutation hypothesis was not, however, fully raised until the 1840s. *Vestiges* was both principal testament and major inspiration for the heated and henceforth public debate of the species problem and its broad and serious moral and social ramifications. *Vestiges* is a bizarre, eclectic, and dogmatic work. Its author sought to trace the development of all things, from cosmic nebulae to animals and man, and assigned the lot to the "sublime simplicity" of law. The Creator was, at most, an indistinct and uninteresting First Cause. In matters of organic evolution *Vestiges* largely reproduced Lamarck. The scientific argument and presumed evidence in the book even sympathetic readers felt outrageous. Hostile readers, like Darwin, found the geology

---

[1] *Zoological philosophy.*

"bad" and the zoology "far worse." The great influence of the widely read *Vestiges* arose not from its scientific merits but from its rigorous and readily comprehensible laying bare of the essential evolutionary issues. The idea of far-reaching change in (organic) nature, a possible if not plausible natural cause therefor, the glaring irrelevance of Scripture for interpreting these phenomena, and the place of man in nature; none of these questions were new and Chambers' discussion contributed little to their resolution. But *Vestiges* provoked examination of these matters by naturalist and nonnaturalist alike. The species problem entered the forum for intelligent and, to be sure, occasionally foolish consideration. There it would remain throughout the century, its notoriety and urgency to be greatly increased and hopes for its resolution to be warmly stimulated by Darwin's *Origin of species*.

On the Origin of Species by Means of Natural Selection

This phrase was the first phrase in the title of Darwin's great work; it was followed by the declaration: "or the preservation of favoured races in the struggle for life." Darwin rightly places the emphasis on the means or mechanism of evolutionary change. He was early persuaded of the fact of such change and it suggested to him that "each species had not been independently created" but had descended "from other species." He recognized, however, that this conclusion would be "unsatisfactory, until it could be shown how the innumerable species inhabiting this world have been modified, so as to acquire that perfection of structure and coadaptation which most justly excites our admiration." The search for this "how" of evolutionary change directed Darwin's consideration of the species problem as it did that of Alfred Russell Wallace. Once delineated, that "how" became the process of natural selection and commanding importance was assigned to the adjective natural. Biological unity in diversity, be it ordinal or temporal, would ostensibly be resolved by complete natural causation.

Darwin (1809–1882) had followed a somewhat casual course of instruction at Edinburgh and Cambridge, seeking first a career in medicine and then in the church. Neither came to fruition. Far more significant in his development were seemingly desultory self-education and intimate contact with distinguished British geologists and naturalists. With them he gained invaluable field experience with plants, animals, and mineral formations. It was above all the Cambridge botanist, John Henslow, who noted Darwin's promise. Not pretending to see in Darwin "a *finished* naturalist," Henslow nevertheless found him "amply qualified for collecting, observing, and noting, anything worthy to be noted in Natural History." Largely on Henslow's recommendation the 22-year-old Darwin became ship's naturalist for the circumnavigation of the globe by the surveying vessel *H.M.S. Beagle.* Impressions and collections gathered on the long voyage of the *Beagle* (1831–1836) induced in Darwin distinctly heterodox opinions regarding the species problem. The years following the return of the *Beagle* were thus decisive for Darwin's intellectual development and by 1844 he had for all practical purposes cast (but not published) his theory of evolution by natural selection.

Darwin himself repeatedly designated the most remarkable "facts" of natural history he encountered while on the *Beagle* and which so provoked his attention upon his return home. They concerned the geographical distribution of South American animals, the relation between extinct and living vertebrates on the same continent and the notable distributional peculiarities of plants and animals on oceanic island groups. This evidence, of course, pertains directly to the fact of evolution and not to its guiding mechanism. For Darwin this fact and the proposed mechanism of natural selection rapidly and deliberately were confused, a confusion so full in fact that, following Darwin's own view, evidence for the former was regarded as confirmatory of the latter.

In Argentina Darwin found two species of a large, flightless bird (*Rhea*) which were separated by no perceptible physical barrier. From this and numerous other examples, he concluded that different but indisputably related forms occur in geographically contiguous areas. Darwin was describing faunal replacement: the occupancy of similar terrains by related but distinct forms as one moves across a major coexistensive land mass. Again in Argentina, gigantic extinct armadillo-like forms lay buried in deposits on the surface of which prospered the modern armadillo; here and in countless other cases appeared that temporal diversification coupled with evident relationship that so greatly impressed all evolutionary biologists. "The mutual affinities of extinct and living species," Darwin later could proclaim, "all fall into one grand natural system; and this fact is at once explained on the principle of descent."

Darwin's most applauded and suggestive evidence came from the flora and fauna of the Galapagos Archipelago. These Eastern Pacific islands, with common geological structure and lying under virtually identical climates, manifested prodigious biological diversity. The kinds of organisms present throughout the islands (tortoises, thrushes, finches, plants) were not numerous but the number and distribution of their species were truly extraordinary. Each island, separated only by about fifty miles of sea from the others, presented a flora and fauna essentially peculiar to itself. Here seemed manifested, as Darwin put it, an astonishing "amount of creative force." That force was unspecified but its effect was unmistakable and at least the possibility that it was a temporal process (and perhaps a natural one) was recorded: "Seeing this gradation and diversity of structure in one, small intimately related group of birds, one might really fancy that from an original paucity of birds in this archipelago, one species had been taken and modified for different ends."

This possibility, we know, clearly formed in Darwin's mind only after his return to England. A justifiably famous entry in his "Journal" reads:

"In July [1837] opened first note book on 'Transmutation of Species' —Had been greatly struck from about Month of previous March on character of S. American fossils—and species on Galapagos Archipelago. These facts origin (especially latter) of all my views."

After filling several notebooks and expending much thought, Darwin in essays of 1842 and 1844 gave full expression to his views. But the essays remained in manuscript form and were known to few. Darwin's attention turned largely to descriptive natural history. It was pressure from Wallace—a singular essay on the geographical distribution of organisms (1855) and the celebrated manuscript announcing a theory of evolution by natural selection (1858)—which returned Darwin to his great theme. The *Origin of species,* an abstract of a proposed larger work, was completed in eight months and published November 24, 1859. The public career of Darwinism soon would begin.

The familiar expression "origin of species" refers principally to the purported transformation of one species into another. Darwin emphasized that he would not consider the ultimate origin of life and hence the aboriginal condition of species, if such there then were. The true starting point and indispensable foundation for Darwin's and Wallace's conception of natural selection was abundant biological variation. Selection has long-term evolutionary meaning only insofar as it designates the act of perpetuating or eliminating differences among organisms. Darwin devoted extraordinary care to accumulating evidences for the omnipresence of variation. His first recourse was to horticulturalists and animal breeders. From them he learned that variation under domestication was the rule and fixed form, function, or behavior the truly rare exception. The varieties of the dog and, more dramatically, those of the domestic pigeon were compelling instances. Turning to variation in nature Darwin, experienced field naturalist that he was, had little difficulty in establishing the same point. His case here was not, however, as persuasive as it had been for domestic variation and it was only in the 1870s that other naturalists, working with museum collections consisting of great numbers of members of a species, firmly founded and extended Darwin's claim for natural variation. The possible cause(s) of this variation, whether domestic or natural, was the object of much uncertain speculation. Here was posed a problem against which innumerable Darwinian biologists would try themselves, for the most part with questionable success.

Evolution by natural selection was founded upon analogy. The great lesson of plant and animal improvement by breeders was that selection would give new direction to the breed. The "key" to these changes, said Darwin, "is man's power of accumulative selection: nature gives successive variations; man adds them up in certain directions useful to him." He then posed, with every hope of a positive response, the decisive question: "Can the principle of selection, which we have seen is so potent in the hands of man, apply in nature?" From the artifice of the breeder Darwin hoped fairly to move to a strictly natural mechanism of change; natural selection was the analogical partner of artificial selection. But the breeder's large hopes and evident skill were based on accumulated experience and perhaps occasional rational expectation. In Darwin's scheme of nature, however, there could be absolutely no place for such intention and design. Anthropomorphism was strictly forbidden. Darwin must find an alternative, a pattern of selection governed by purely natural factors. His search for such an agency led to exalted appreciation for the role of competition in the general economy of organic nature and to his well-known exploitation of Thomas Malthus' consideration of the factors limiting the size of human populations.

That plants and animals were perpetually engaged in severe competition with one another and with their environment was already a familiar argument in Darwin's youth. Such competition was usually viewed as primarily destructive, that is, when severe it led to the elimination of the less adaptable or less prolific forms. Those of pious or sanguine temperament viewed this process as an admirable provision by the Creator for ensuring the equilibrium and cleanliness of His creation.

Malthus' vision was cast in a more somber setting, the horrors of over-population and wretched death induced largely by the rapid and undirected urbanization of Britain during the Industrial Revolution. Malthus, too, derived his laws from that "Being who first arranged the system of the universe." Times, however, were hard and Malthus saw no basis for future improvement. Population increases geometrically; food, indispensable for life and prosperity, augments but arithmetically. Checks on population must therefore act, and their action was remorseless. "Necessity," Malthus announced,

"that imperious all-pervading law of nature, restrains [life] within the prescribed bounds. The race of plants, and the race of animals shrink under this great restrictive law. And the race of man cannot, by any efforts of reason, escape it."

Plants and animals perish from "waste of seed, sickness, and premature death;" man expires from misery (famine, disease and war) and vice.

This thorough-going pessimism poses a large problem for understanding the enunciation and bearing of the argument from competition as given by Darwin. The naturalist later recalled that in October 1838, he read Malthus' *Essay on Population* and was struck that under conditions of struggle "favourable variations would tend to be preserved, and unfavourable ones to be destroyed." A "new species" would result; Darwin "had at last got a theory by which to work." But had he? In a strict sense, it seems that he had not. Rigorous Malthusianism is contentedly pessimistic; its outlook implies that the only permissible change is destructive change; our world is severe and closed and as such the product of the divine Legislator. Whether evolutionary change be defined as simple transformation without reference to direction of change or as progressive transformation, with man as the presumed standard of progress, the Malthusian argument cannot be of decisive contributory importance. What Malthus brought to Darwin (and to Wallace in 1858) was a renewed sense of the powerful, indeed overwhelming, place of competition in the organic world. Malthus' scheme obviously assigned no place to development. It would be Darwin's task as he worked toward a delineation of evolutionary natural selection to introduce just this notion of temporal change that leads to, and consequently explains, organic diversity and relationship. Given, then, the fact of variation—to be observed especially from the viewpoint of the differential propensity for survival under given ambient conditions—and that of unceasing competition, the process called natural selection was laid bare: "This preservation of favourable variations and the rejection of injurious variations, I call Natural Selection." Characters of neutral value or those whose usefulness or deleteriousness was not apparent would remain untouched by selection. Darwinian selection derives its operative power from the postulate of utility, that is, survival, and hence evolutionary change is strictly a function of the (presumed) structural or physiological value of every organic part or process. The then rampant utilitarianism of political economy and moral philosophy reached into all areas of thought, and not least Darwin's notion of survival value.

But what could natural selection do? It was responsible, above all, for "divergence" of character and for extinction. Divergence was to Darwin the critical moment in evolutionary change. Here it was that the multitudinous offspring of a given species broke up into reasonably well-marked "varieties" and, should the pressure of selection not abate and the flow of time proceed, these varieties would in turn be transformed into veritable new "species." Natural selection would accumulate small variations and bring forth major transformations. Each variety and each new species would ever capitalize upon the physical and biological environment in which it might find itself. As a consequence, Darwin took it as a principle that the "greatest amount of life can be supported by great diversification of structure" and function. The Malthusian analysis pointed to the incessant pressure to increase the "amount of life." Darwin accepted this analysis, explaining by it extinction and, the great triumph and originality of his theory, the emergence of biological novelty. This was evolutionary change and it meant primarily the radiation of life into every available niche in the environment. Evolution was totally opportunistic and organisms benefited from whatever advantage variation might confer upon them. Natural selection became thereby an exceptionally plausible explanation of adaptation, revolutionizing one's view of apparent (creative) design in organic nature, and also of the familiar, perplexing, and fundamental relatedness of great and diverse groups of organisms. The "element of descent," Darwin proudly announced, "is the hidden bond of connexion which naturalists have sought under the term Natural System" of classification. Genetic explanation is here exhibited in full colors.

It must be noted briefly that Darwin used the term "struggle for existence" only in a "large and metaphorical sense." It determined not only "the life of the individual, but success in leaving progeny." The distinction is significant, since Darwin, admitting a fair degree of

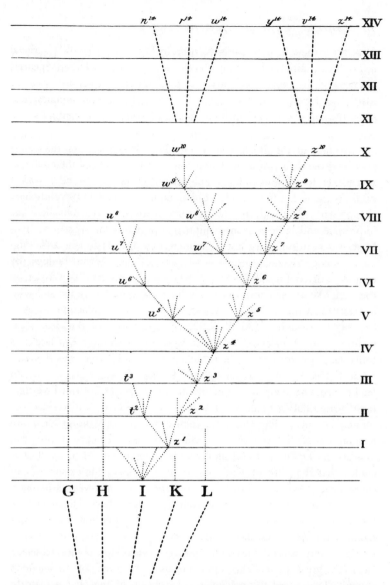

Figure 4.3 Evolutionary change on the Darwinian model is a cumulative process. The selective addition of slight and favorable variations leads to the divergence of species. In this diagram Darwin traced, hypothetically to be sure, the offspring of single species (I) over 14 thousand generations. He hoped to demonstrate the possibility of deriving through descent with modification at least six quite distinct contemporary species and to expose the no less important and common fact of extinction of species (G, H, t^3, u^8). (Charles Darwin, 1860.)

ferocity in the life of organisms, nevertheless felt differential survival to be "more important." Many Darwinians, including Herbert Spencer, who coined the expression "survival of the fittest," also emphasized the gladitorial aspect of survival and did so largely for nonbiological reasons. Darwin chose to keep both aspects of competition continually in play.

Only four of the fifteen chapters in the *Origin* were devoted to expounding the theory of natural selection. The remainder of the book was given over to consideration of difficulties of the theory and a review of the various kinds of evidence that both demonstrated evolutionary change and confirmed the new theory. Darwin's reflections here are impressive and largely defined the major provinces of subsequent biological Darwinism. They testify, moreover, to his prodigious and long-pursued assembly and mastery of all facets of evolutionary biology. By contrast, Alfred Russell Wallace (1823–1913) appears a newcomer on the scene. Wallace, a man of extraordinary personal character and with interests far broader than Darwin's, was self-taught. Sometime in 1844 or 1845 he read Lyell, Malthus, Darwin's narrative of the *Beagle* expedition, and *Vestiges*. The effect was overwhelming; Wallace was converted, or corrupted, to the evolutionary faith. *Vestiges* was the critical text. His mission was now clear—he had to find facts that would lead to "solving the problem of the origin of species." From investigations carried out in Brazil and the Malay Archipelago Wallace made of himself an outstanding tropical naturalist and perhaps the greatest biogeographer of the century. Reflecting on his self-imposed problem, Wallace in February 1858, and wholly independent of Darwin concluded that the "life of wild animals is a struggle for existence." In consequence, "we have *progression and continued divergence*" of organisms and this can be "deduced from the general laws which regulate the existence of animals in a state of nature." Competition and differential survival were these laws. The species question was again resolved by exclusively natural causation. Wallace had all too obviously seized Darwin's great prize. The simultaneous publication of their views (July 1, 1858) and the precipitate preparation of the *Origin* seem to disclose, however, unusual and genuine harmony between the men. Priority disputes did not arise and, while Darwin reaped the greatest share of public attention, Wallace became the impassioned, indeed doctrinaire, British advocate for the all-sufficiency of natural selection as an evolutionary mechanism. Darwin himself would often waver under the frequently devastating criticism which soon was levelled against the novel doctrine.

The theory of natural selection radically and definitively altered discussion of the species problem. Only keen enthusiasts pretended that selection was more than a highly probable explanation of species transformation. But that probability was sufficient: it provided a coherent and open-ended alternative to the traditional creationist view of species and their interrelationships, distribution and adaptations, as being the product of supernatural concern and action. Just as uniformitarianism had banished geological catastrophes from serious scientific consideration so, claimed Darwin, will "natural selection, if it be true, banish the belief of the continued creation of new organic beings."

Having found a plausible evolutionary mechanism meant, moreover, that evolutionary change could no longer be smartly rejected because one might be unable to envisage how that change had occurred. The fact of evolution thereby gained new importance and evidences favoring it now received comprehensive and generally fruitful attention. The question of whether selection had really "acted in nature" Darwin believed "must be judged by the general tenour and balance" of this evidence. However this be, and Darwin's claim may well be disputed, natural selection opened the way to a new era in evolutionary studies. Evolution became the unifying theme in botany and zoology, binding together in the common pursuit of the actual history of life on the earth the special disciplines of classification, paleontology, comparative anatomy, embryology, and ecology. Darwinism overwhelmed biology and, as the fashion of the age, not surprisingly lent its prestige to realms of inquiry to which its relevance was highly debatable.

Varieties of Darwinism: Biology

The impact of the *Origin* upon biological thinking was immediate and dramatic. By no means did all biologists accept natural selection as a fully competent mechanism of evolutionary change; Darwin's theory would experience many vicissitudes before its firm integration into biological explanation during the 1930s. The idea of descent with modification, that is, the fact of evolution, met with notably less scepticism. Indeed, one might define biological Darwinism as the general acceptance of this fact and, more importantly, the widespread effort after 1859 to reconstruct so far as was possible the actual course of past evolutionary transformations of life. An era opened in which traditional studies were recast and applied anew to the recovery of a past which once had been denied any existence whatsoever and was still regarded by sceptics as beyond the reach of scientific inquiry.

In no discipline did expectations appear so great but frustrations prove so common as in paleontology. Darwin and the Darwinians all recognized that the fossil record, the very product of past life on the earth, should provide the truest foundation for the desired reconstruction. But the fossil record was woefully incomplete. In part this was a result of simple inadequacies of collection and study. Fossilization was, however, a complex and often capricious matter; perhaps one had no right to expect complete fossil series representing the emergence of various organisms. The strata in which fossils occurred were, moreover, commonly disturbed and successive formations were separated by demonstrable and serious discontinuities.

Paleontology as a consequence long maintained most ambiguous relations with orthodox Darwinism. The majority of paleontologists felt their evidence simply contradicted Darwin's stress on minute, slow, and cumulative changes leading to species transformation. Schools of saltatory evolution, advocating species change by sudden and systematic alteration of form and function, developed. Other paleontologists, usually from an older generation, continued to reject evolutionary change.

Darwin nevertheless found amply qualified advocates. Thomas Huxley in Britain, Alfred Gaudry in France, Melchior Neumayr in Germany, Vladimer Kowalevsky in Russia, and Othniel Marsh in the United States all worked diligently, collecting important although not truly complete fossil series and reading their evidence as testimony for evolution. Their objective was clearly stated (1865) by the eminent Swiss paleontologist, Ludwig Rütimeyer: paleontology and zoology "must in the fullest sense of the word be the *History of Nature* and trace out the threads which bind together the present and earlier generations of organisms." In some cases (corals, ammonites, horses) reasonably distinct lines of descent had been recovered by the end of the century. This evidence was strongly suggestive, if not decisive, and paleontology since 1900 has expanded it both in substance and comprehensiveness.

More ardent evolutionary reconstructionists found some encouragement in morphology. Scrupulous attention to anatomical comparison would reveal, many believed, "the relations of affinity within the various divisions of the Animal Kingdom." That affinity was nothing less than the genetic bond that descent with modification necessarily imposed as organic nature became diversified. It was from embryology, however, and not anatomy that the greatest rewards were expected. The recapitulation hypothesis, already seen developing and under attack early in the century, returned after 1859 and pursued a truly extravagant existence

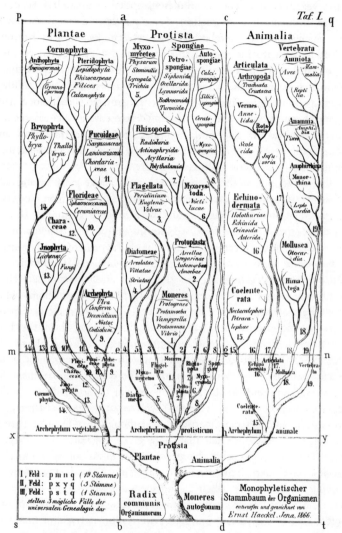

Figure 4.4   *Post-Darwinian biologists relentlessly pursued recovery of the history of life on the surface of the earth. To concrete studies in morphology and paleontology was added both shrewd and fanciful speculation. This elegant phylogenetic tree, perhaps the first to appear after the publication of the* Origin of species, *offers three possibilities (19, 3, or 1 "stems" or primitive forms), the author favoring the last, for the much desired "universal genealogy" of all living beings. (Ernst Haeckel, 1866.)*

for several decades. The recapitulation episode offers a classic instance of the fashions that so frequently and pervasively sweep the sciences and doubtlessly most intellectual pursuits. The remarkable parallel between individual development and ancestral history appeared to present a powerful tool for recovering that past which paleontology regrettably could not seize. Turn to the embryo, it was ordered, and follow its every permutation. There, under the microscope, was laid bare the ancestral condition of life; uninterrupted attention to such development offered the observer a veritably complete record of the course of evolution. Comparative embryologists insistently urged the value of the recapitulation hypothesis. The coming, great generation of experimental biologists, trained under these men largely between 1870 and 1890, began as students of evolutionary reconstruction. It was, of course, largely they and not their mentors who brought forth the facts and argument that greatly discredited the claims of a predominantly embryological route to exact reconstruction of the past.

However, these embryological investigations served enormously to expand biologists' knowledge of the varied developmental patterns of plants and animals. This knowledge had practical value, offering, for example, more trustworthy and often new foundations for classificatory work. In the case of those animals, common among the invertebrates, exhibiting alternation of generations, mere inspection of different forms had led to the singular conclusion that different stages of the "same" species were different species altogether. The quality of "sameness" was, of course, quite unsuspected. Only by patient observation of the whole reproductive, indeed life, cycles of such creatures was it discovered that these differences, extraordinary and real though they be, were but part of a definite developmental plan of a given species. Embryology here virtually proved, by tracing change, that "sameness" on which a reliable species description might be based. At century's end embryology had become but one guide, albeit an indispensable one, to the broader pattern of life's history; paleontology was no less. With no single or omnipotent clue to the past, evolutionists were ultimately compelled to disdain fashion and exploit evidence wherever it might be found.

Biological Darwinism supported abundant critics and their observations were rarely confined to the problems of evolutionary reconstruction. No aspect of Darwinism is more confusing, and has been less studied, than the biologists' inconclusive and tortuous discussion of the efficacy and failings of natural selection. In large part this conclusion was a consequence of the very imperfect knowledge then available regarding biological variation. It was no less due to the visceral aversion on the part of many able biologists to orthodox Darwinian reliance upon utility.

Descent with modification proceeded, according to this latter view, by the exceedingly slow accumulation of slight variations. Those creatures that survived did so because the aggregate of their several parts had proved useful. Every part at all times was subject to the severe review of selection, and only those variants that ensured survival or net reproductive success would in the long run persist. The critics charged that Darwin had offered only a verbal solution to the problem, for what was utility and how was it to be measured? Certainly no absolute and truly general standard of usefulness could be devised. The claim that survival itself was the test of fitness and hence of utility merely reopened the famous charge that selection theory was a tautology.

Worst of all, what conceivable use could these indispensable and extremely minute variations present at their first appearance? In the struggle for existence they could offer only trivial advantage to their possessor; they would be immediately "swamped" in the great numbers of a breeding population and thus lose all evolutionary meaningfulness. Major critics of this temper customarily turned to saltationist alterna-

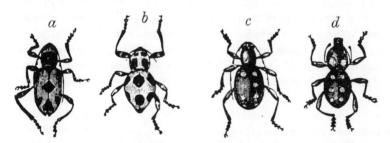

Figure 4.5   Protective coloration—the evolutionary acquisition of the external appearance of a noxious form by another group of organisms more apt to serve as prey—was discovered in the 1860s and interpreted in Darwinian terms. Selection favored those forms whose variations made them most resemble or mimic the undesirable form; truly remarkable cases of resemblance were described. In this case, soft and hence more edible genera (a, c) of Philippine Island beetles mimic the hue and pattern of coloration of the hard-covered and hence protected genera (b, d) of weevils. (Alfred Russell Wallace, 1889.)

tives. Assigning to each organism "its own internal and special laws of growth and development" they made these laws responsible for the appearance of all novelties; selection at best could secondarily pick and choose from among the macrovariations then presented. Despite the fact that such "internal laws" were patently *ad hoc* inventions they offered the reassuring illusion of having resolved the problem of the cause of variation and of having reduced the creative significance of radical selection theory. Not until 1930 would the problem be successfully translated, through rigorous statistical analysis of the survival value of minute variations, from the realm of varying individuals to that of the variations distributed in large breeding populations. The coming then of population genetics permitted the definition of a neo-Darwinian school of selection theory.

Biological variation in the post-Darwinian era was itself a widely recognized fact whose potential resolution into cellular phenomena was a no less widely accepted faith. That faith was based more on the closure of other physiological alternatives than on decisive, confirmatory evidence. On the one hand, the years 1860 to 1900 witnessed an astounding proliferation of speculative theories of inheritance; on the other hand, it was at this time that, in general isolation from one another, the diverse disciplines that would give rise to the science of genetics matured. Cytology, experimental plant and animal breeding, and the application of statistics to biological problems came to the fore during these years. With the rediscovery in 1900 of Mendel's experimental data and the broadcasting of his conclusions, previously disparate studies came together. Their disciples became geneticists and their work established the particulate basis, first in the chromosomes and then in the hypothetical gene, of heredity and variation (see Chapter III).

By 1900 biological Darwinism had overwhelmingly prevailed. While criticism of natural selection remained common and was often damaging, no serious alternative, other than the ever-dubious environmental (called neo-Lamarckian) hypotheses, could be seriously maintained. The fact of descent with modification had become the permanent acquisition of all biologists. Said an outstanding textbook of the 1880s:

"The origin of plants and animals by descent from other preexisting forms can no longer be questioned; and it is evident that the origin of any given plant or animal has been a definite process, to be determined like any other facet of natural history."

In keeping with this conclusion, biologists sought to recover both the general pattern and intimate detail of the history of life. In the former endeavor they largely succeeded; in the latter, their successes were real but by no means complete. This impassioned quest for phylogenetic reconstruction was as much a gift to twentieth-century biology as the vexing issues surrounding the nature, causes, and consequences of biological variation and the general question of the evolutionary mechanism.

Varieties of Darwinism: Theology

The leading Darwinians were determined to place man in nature and make him an integral part and indubitable product of nature. While their objective was commonplace, their success was not, and that success, by lending to their endeavor all the prestige and ascribed certainty of biology as a science, did not fail to reheat already warm disputes between the secular man of science and the supporter (a category including many distinguished scientists) of the Christian conception of man. The dispute was truly joined over moral matters and not the interesting but lesser issue of man's physical relationship with the animals. "Man," the apologist insisted, "is endowed with a moral nature, a perception of right and wrong, and a feeling of moral obligation." This moral sense was man's alone, a cherished endowment his by will of the Creator. It determined his character and specified the all-important relationship of man and his God. Man, the Christian believed, carried with him sin from the Fall. His only hope for salvation from eternal torment lay in his acceptance of Christ and the power of Atonement. Man morally was not self-sufficient; he needed Christ or at least the church.

It is not difficult, therefore, to appreciate the horror with which the concerned theologian would greet the following view, expressed by Darwin in 1871 but by then familiar coinage, of the origin of man's moral sense.

"Ultimately a highly complex sentiment, having its first origin in the social instincts, largely guided by the approbation of our fellow-men, ruled by reason, self-interest, and in later times by deep religious feelings, confirmed by instruction and habit, all combined, constitute our moral sense or conscience."

Quite apart from this implicit and inadmissible moral relativism, apologists believed the evolutionist promulgated a doctrine devastating at once to morality, church, and society. Man was no longer a fallen creature whose fate only Christian belief or the church could decide. He was, rather, an independent being, produced by nature and wholly divorced from God. In the dissolution of the eternal sanctions of morality man seemed to become no more than a highly complex and rational social animal. All religious mystery surrounding man's origin or destiny could easily, fervent atheists had amply shown, be shed.

The charge of atheism was commonly brought against the Darwinians. It was probably an unwarranted label, for most evolutionists—Haeckel was an outspoken and consequently malodorous exception—were too firmly bound by reason or sentiment to the traditional values of their society to encourage so great and irrevocable a break with the past. Darwin very properly concluded that his "theology [was] a simple muddle;" Thomas Henry Huxley retired into agnosticism; Lyell remained true to his faith. Most evolutionists, in short, did not directly campaign against Christian teachings, even though they did not fully share them and constantly declared their outrage at the pretentions of learned churchmen.

To be sure, the theological response to Darwinism involved issues other than the moral independence and responsibility of man. Biblical literalism, including the still prosperous Design argument, bent only slowly before the combined onslaught of geology and evolution and the rapidly advancing historical and critical study of biblical texts. The Roman Church, doctrinally increasingly conservative toward the end of the century, seemed more alarmed by modernist interpretations of Scripture than interested in the evolutionists' disputes. The more fundamentalist Protestant sects preserved their Bible inviolate. Other Protestant sects, relying heavily upon allegorical understanding of the Mosaic creation account, sought to break away from the embarrassing confines of six literal days of creation.

The task of reconciliation between evolutionary biology and Christian doctrine suggests both an original, if unarticulated, union between them and subsequent rupture. The latter was one of the nineteenth century's most notoriously public facts. However, the former was, in the minds of many, merely another manifestation of the theologian's presumption to grasp all knowledge as intrinsically his and with divine bearing. Here were ripe grounds for personal and intellectual tensions. The possibility for the joining of dispute could hardly be avoided once the evolutionist made clear his naturalistic conception of man's purportedly highest qualities and whole being. Evolution thus provided a prominent battle-ground for the ongoing contention between science and religion for the allegiance of the European mind.

Varieties of Darwinism: Society and Race

Of the many varieties of Darwinism social Darwinism is easily the most familiar. It is also the least amenable to exact characterization. Social Darwinists were a diverse lot and their premises and objectives on the whole constitute no consistent doctrine. One may, however, isolate leading, if not in every instance predominant, themes in social Darwinism. Certain of these themes, of course, antedate Darwin and were first cast in a distinctly nonbiological context. Here as elsewhere a major contribution of evolutionary thought was substantive and also the presentation of an ostensible scientific grounding for views already current and soon to be revivified and revised.

Darwinian selection theory founded evolutionary change on the accumulation of numerous slight variations; the prime entity was the varying individual. With some qualification this was a view shared by classical liberal political economy. (Darwin's ideas are often attributed to this source). The accent fell heavily on the determinative value of the individual, each of which sought unrelentingly its optimum advantage. In a human social situation this fact announced the existence of conflicting interests, hence, competition. Success in the rivalry could have diverse meanings: profit, enhanced social status, political or military power; indeed, if one were thus inclined, it could and did mean sheer life and death. As noted, biological Darwinism shared and drew upon this argument; no less did it shed its aura of scientific credibility upon it. Categorical individualism and competition ensuring some recognizably desirable advantage to the successful were essential to the dogma of social Darwinism. Natural selection seemed equally effective in nature and society.

The metaphor of the social organism also entered most social Darwinian discourse. This ancient metaphor, recoined for the nineteenth century by German historians and nature-philosophers, presented rival alternatives to the social Darwinian. He could, on the one hand, emphasize the integrity and delicate balance of the organism. Such a view surely forbade all efforts at major or sudden reform of social conditions; this version of the social organism was a call to gradualism and noninterference with nature's prerogatives. The social reformer could, however, choose to call on the well-known adaptability of organisms. He stressed the organisms's responsiveness to changed environmental conditions and founded his hopes for social amelioration on man's capacity for understanding and then controlling this environment. Both positions represent fairly social Darwinism but, while both had their numerous and vocal partisans, social Darwinism was in general a conservative movement.

Any discussion of social Darwinism leads directly to consideration of the development and delimitation of the sciences of man. The role of Herbert Spencer, a leading social Darwinian, and his relation to decisive currents of continental social thought will be examined in the following chapter. Two examples will illustrate here trends of thought in conventional social Darwinism. They come from the United States and Prussia, the former asserting its manifest destinies following the disunity of the Civil War and the latter simultaneously embarking on the final and successful drive to make a unitary nation of the long-fragmented Germanies. From about 1860 both states entered an era of rapid and generally rapacious industrialization. Wealth, power, and status were to be won; they would necessarily go to the strong. It seems hardly fortuitous that the United States from 1865 and the German Empire somewhat later nurtured the most ebullient and harsh social Darwinians and that here especially that doctrine offered welcome credibility to the advancing arguments of racism.

William Graham Sumner (1840–1910) was satisfied to develop only the intrinsic and valuable competitive basis of social organization. The critical factor was the ratio of man to land, land being the ultimate source of wealth. As the value of this ratio increased, so heightened competitive struggle. Social fitness would be measured by money: "millionaries are a product of natural selection, acting on the whole body of men to pick out those who can meet the requirements of certain work to be done." Under severe competitive conditions democracy was undesirable and wrong. All efforts at social action to relieve the difficulties and suffering of the masses were equally misguided; society was an organism and would change only at a (slow) pace which it, not man, imposed. Sumner, first a preacher, then a Yale sociology professor, perpetuated traditional Protestant values of hard work and frugality and turned indignantly away from eighteenth-century American illusions of egalitarianism. Sumner's objective, premises, and argument were sociological; they borrowed terminology, modernity, and seeming substance from Darwinism.

German thought was only thoroughly infected by social Darwinism after 1890. It encountered an already lively tradition, the idea of a German *Volk,* an idealized, physically and spiritually bound group or "people" whose unity and supreme merit were indubitable, unique, and closely associated with the German countryside. The idea of the *Volk* and social Darwinian themes combined to produce a racial consciousness that ranged from caution to proudly brutal fanaticism.

Alfred Ploetz (1860–1940) illustrates the moderate basis on which subsequent Nazi racial barbarism was able to draw and build. Ploetz, who died a professor in Hitler's Berlin, began in 1895 to create a science of racial hygiene. His approach was, in essential contrast to many of his successors, generally humane. He advocated racial improvement by means of proper mate selection; controlled breeding would, if socially enforced, gradually lead to a more healthy and racially desirable nation. His plan nevertheless included review by a panel of experts of newborn babies and the decision as to whether the child was racially profitable or was to be forthwith eliminated. To Ploetz, the "race" was all mankind. To other and particularly to later National Socialist theorists this view was sentimental and too ecumenical. There existed many and distinct races, of which the Aryan or Nordic was best and the Hebrew obviously the worst. To the ardent racist or Nazi mere regulation of breeding would hardly suffice. Stern measures were in order: sterilization; social, legal, and economic segregation; and, the infamous final solution, mass racial extermination.

Certainly social Darwinism need not necessarily have contributed to such deplorable conclusions and actions. It cannot be denied, however, that the socio-biological pseudoscience which largely was social Darwinism appealed to its proponents not only because of its scientific tone but because it offered justification for ruthless individual or social action. Many in the United States called upon and continue to call upon biology to remedy or explain away the endemic racism or ethnocentrism, be it directed against Italian or Indian, Jew or Negro, so pronounced in this country. Virulent racism, of course, introduced value judgments that are not derived from the scientific study of biological differences. The temptation to do so has nevertheless always been present and was greatly stimulated by the extension of Darwinian evolutionary themes to social analysis. Responsible biologists have long deplored this practice and continue today to find it necessary to offer sane consideration of the biological meaning of race and to attempt to contravene unwarranted misapplication of their conclusions.

Transformations

The coming of evolutionary theory in biology offered an alternative to the traditional Christian account of creation. The creationist and the evolutionist shared an interest in organic adaptations and diversification of species. Their manner of explaining these prime facts of scientific natural history reveals, however, the fundamental difference in their outlook and aspirations. The one assigned causal efficacy to a supernatural being or will, the other attempted to account for the phenomena without reaching beyond natural, hence experiential, means. A semblance of compromise was always possible; it usually took the form of assigning phenomena to natural laws which were in turn the direct product of creative wisdom and power. This would not, reasonably enough, satisfy those who sought purely naturalistic explanation.

Darwin and Wallace had shown that organic transformations were the rule, not the exception, in nature. They and their party supported their conclusion by bringing to bear evidence that exceeded, both in quantity and pertinence, all that had previously been gathered. They drew upon this earlier material but did so for a clear-cut purpose, to document the course of evolution. That same purpose dictated no less the direction of their own original investigations. Darwin and Wallace furthermore explored the possible explanations of organic change and settled, of course, on natural selection.

Within biology the triumph of evolution by natural selection presented unanticipated opportunities. Evolutionary descent served as a focus about which developing biological specialties gained orientation and won repute. Rare would be the paleontologist, anatomist, or embryologist of the 1870s who had not related his work to evolutionary doctrine or failed to assume a distinctly favorable (or unfavorable) stance with regard to the Darwin-Wallace notion of natural selection. Descriptive and classificatory biology of the later nineteenth century were thoroughly permeated and their objectives and practice largely defined by evolutionary themes. Physiologists, however, were more divided over the merits of Darwin's proposals and often highly doubtful of their relevance to functional biology.

In a century of science the success of Darwinism could scarcely fail to attract attention. Since science was commonly believed to offer consummate certainty to the understanding, it was probably inevitable that Darwinism would be pillaged for moral, social, political, and all other manner of nonbiological conclusions. If man, after all, were an animal then certainly he and his society should be subject to the conclusions developed by the science of animals. Simplistic though it may appear —and in reality be—this argument gave credence and vigor to a great deal of the popular social Darwinizing. It is not surprising, therefore, to see the elaboration toward 1900 of a self-sufficient science of man valiantly attempt to root out the obscurities and errors that it viewed as the heritage from Darwinism.

CHAPTER V

Man

THE INITIATION OF THE SCIENTIFIC STUDY of man is often associated with the coming of Darwinism. Certainly this view was proclaimed by the newly defined evolutionary biologists themselves. Thomas Huxley in 1863 announced that "the question of questions for mankind—the problem which underlies all others, and is more deeply interesting than any other—is the ascertainment of the place which Man occupies in nature." Darwin in the *Origin of species* had cautiously suggested that by his hypothesis of descent with modification "light will be thrown on the origin of man and his history." Forty years earlier the first systematic transformist, Jean Baptiste de Lamarck, had argued that man, a "true product of nature, the ultimate measure of the eminent products which [Nature] can bring forth on this globe, is a living body within the animal kingdom and belonging to the class of mammals." And forty years after publication of the *Origin,* Ernst Haeckel assertively claimed that "the descent of man from an extinct Tertiary series of Primates is not a vague hypothesis, but an historical fact."

Between Lamarck and Haeckel a revolution in man's awareness of his past had indeed occurred. Although Lamarck spoke principally from the implications of the first principles of his science, Haeckel, as well as Huxley and numerous other post-Darwinian students of man, could cite sound evidence for their claim that man was at once the product and an integral part of nature. Their evidence derived from comparative anatomy, fossil human remains, and the traces of prehistoric societies. Although incomplete and in part open to serious question, this evidence was highly suggestive. It gave birth during the nineteenth century to several remarkable specialties, among them physical anthropology, human paleontology, and archeology. Each of these new disciplines was historical in outlook and concentrated on the physical relationships and remains of the human frame or the concrete artifacts of his various cultures. In a general sense physical anthropologists fastened their regard on individual specimens and the archeologists attended most closely to the traits of particular past cultural situations. But the historical element, introduced, as will be seen, from diverse sources, compelled always a wider view. One's interest in man might well be more comprehensive than the structure or artifacts of individual men. One could and frequently did take all mankind as the object of inquiry.

Herewith opened a quite different perspective on man. The nineteenth century, following ancient concerns forcefully recast during the eighteenth century, sought to recapture the development of the human mind. Emphasis might fall as easily toward man the moral being as to the physical aspects of man's animality. "Moral" is a conveniently ambiguous expression. In a narrow and conventional sense it referred to the essence of man as seen through Christian eyes. The naturalistic description of man set in motion by the transformist school and notably the Darwinians thus appeared wholly to destroy the moral being of man (see Chapter IV). But moral referred also to attributes pertaining to the mind and character of man, and thus to what was commonly received as the incomparable, truest, and highest human quality, reason. In this sense will be understood the "moral philosophy" of the eighteenth and nineteenth centuries, a moral philosophy whose premises profoundly influenced the developing sciences of man.

Eighteenth-century moral philosophers postulated the psychological unity of mankind. The mental vagaries of individual men were consequently distinctly less interesting and important than those properties of mind shared by all men. The student's concern thus concentrated on man as a collective or social being. The very object of such study was man in society or, simply, society. Particularities, be they of mind, body, or cultural setting, assumed significance only insofar as they contributed to our understanding of the overall development of mankind. The historical factor continued, of course, to offer explanatory satisfaction but the pattern of that history was now seen as universal, encompassing societies of all times and places, and not as the confined and minute record of prominent persons and individual events.

One must not assume that the protagonists of the individual and of the social nature of man and his works spoke in necessary opposition to one another or that the two approaches were mutually exclusive in any given investigator. It is, however, a fact that the nineteenth century increasingly turned its regard toward social phenomena and discovered there problems urgently requiring resolution and suggestions for devising a dependable science or sciences of society. Man was necessarily the object of this new science as he was no less the object for the Darwinian's analysis. While the last third of the century marks the birth-period of self-conscious and assertive sciences of man (psychology, anthropology, and sociology), they derive primarily from the moral philosophy and sociopolitical speculation of the years prior to 1859 and owe very little in terms of explicit content to the triumph of Darwinism. The popular conclusion that the scientific study of man depended largely on the successes of evolutionary biology requires, therefore, critical inspection and serious qualification.

Man: A Created Unity and the Fact of Diversity

The study of man as related here concerns exclusively the Christian world. Christian orthodoxy required the unity of mankind. God the creator had formed Adam, from him made Eve, and from this original pair descended all humanity. The scriptural account was powerfully reinforced by the arguments of Saint Augustine. To Augustine man was necessarily one and the peoples of Europe and her perimeter, despite their apparent differences, were the descendants of God's first created pair (monogenism). Most importantly, Christ had come but once and then for the salvation of all men. The promise of Christ's life, crucifixion, and resurrection was simply too great to be closed to any man. Augustine's view, suggestive of later medieval thought on the relationship of human forms and societies, was necessarily confined to the small European world.

But Europe of the Renaissance was expansive. An era of worldwide exploration was opening. Medieval travellers had shown remarkably little interest in the strange peoples they encountered. Travellers after about 1500, be they in search of wealth, territory, adventure or souls-to-be-saved, were deeply impressed by the peculiarity and multiplicity of the new peoples they encountered. The human body seemed cast in countless distinctive forms; the customs and manners of society, from dress to dwellings, from family organization to religious practice, from speech to modes of thought, appeared varied beyond the wildest reaches of human fancy. An astonished Christian Europe was nervously rediscovering human diversity.

Who were these people? What connections, now presumably lost in history, had they maintained with Europe? The latter question is critical, for it confronts the scripturally necessary unity of mankind with the manifest and surprising facts of human diversity. One could premise an aboriginal diversity of mankind (polygenism), in so doing, of course, paying minor heed to the Mosaic creation account. This was the route to heresy and it was, while not uncommon from the late Renaissance through the nineteenth century, distinctly less desirable than some accommodation with scriptural needs. Two principal alternatives were available. One might scrutinize *Genesis* anew and find therein unsuspected sources of human diversity. Much was made, for example, of the Ten Lost Tribes of Israel; they were cast as the possible ancestors of the Ethiopian, the English, or the American Indian. Cain's offspring, unrecorded in scripture, might well have populated the globe with strange tribes. And whence came Cain's wife, a partner whose obscure parentage was a constant embarrassment to literal readers of *Genesis*?

A second alternative, and one rich in associations with the life sciences, postulated an orthodox unitary creation followed by the incessant modifying action of the environment. Unity could thus be preserved and diversity no less accepted; the one was primary and the other a secondary, perhaps explicable, natural production. Environmentalism was known and advocated by ancient Greek students of man, most notably by the Hippocratic physicians and Aristotle. It disappeared in the Christian West until the classical revival of the Renaissance and then returned, largely through the efforts of political theorists, to capture a secure place in European reflections on the constitution and distribution of mankind. Environmental influences acting upon the living organism, animal or human, played of course a leading role in the doctrines of the early transformists, most notably in the speculative system of Lamarck.

The necessity of reconciling human diversity and the scriptural need for unity had not diminished by the nineteenth century. The explanations offered continued to be strikingly familiar. Polygenism revived in uncommonly stern form in the writings (1830–1860) of American apologists of racism and slavery. Environmentalism expanded into virtually all aspects of biology that dealt with the transformation of organisms, including man. Scriptural reinterpretation, embracing special concern for the Mosaic account of the creation, was among the century's most vigorous and distinctive intellectual occupations. Not unexpectedly, the Darwinians felt they brought a fresh solution to these problems. The matter of origins was of lesser interest to them. The "chief philosophical objection to Adam," Huxley announced, was "not his oneness, but the hypothesis of his special creation." The Darwinians believed that, unlike Lamarck's clearly faulty hypothesis, their mechanism for evolutionary change (natural selection) would contribute both system and scientific credence to the study of organic transformations. The diversity of man no less than that of animals and plants could thus be satisfactorily explained. This diversity was naturally produced and one could be quite indifferent, as Huxley's words reveal, to the aboriginal condition of man. Diversity could be produced from either aboriginal oneness or multiplicity; the hypothesis of a special creation was unnecessary, lacked corroborating evidence and limited without justification the scope of evolutionary inquiry.

Man's Place in Nature

The Darwinians confidently assumed the animality of man and the legitimation by this fact of the applicability of the methods and conceptual system of biology to the study of man. That man is an animal was no discovery of the nineteenth century. This conclusion was common in classical antiquity and, with suitable reservations, found a home in Christian orthodoxy. The critical reservation demanded man's unique possession of a soul. Aristotle had granted man, in addition to the nutritive and sensitive "souls" he shared with animals, an intellective or rational soul. On the highest level of his being, then, man already was set apart from the remainder of creation. Christian doctrine, while of course attributing to soul a new and peculiar meaning, vigorously supported the same discrimination. Man, unlike all other creatures, was endowed by the creator with a soul. This possession made man alone a free and moral being. The spiritual aspect of man was obviously being elevated well above his physical constitution and relationship to other creatures. Here was a systematic and persistent bias exhibited by all who chose to study man, a body of thinkers who, as has been seen, were deeply committed to oneness of mankind.

This bias began to break down only in the late seventeenth century and appears to have followed closely upon the recognition of the real physical (and cultural) differences that separate men. Writing in the 1670s, an Oxford anatomist and distinguished political economist, William Petty (1623–1687), exploited the common idea of the great chain of being with the aim of seeking both the differences between man and certain animals and the distinctive gradations among men themselves.

For this purpose only salient and dependable characters might be employed and these characters were to be apprehended only by the senses. We must rank the varieties of mankind insofar as they are "like in shape and visible appearance, rather than by the Objects of other Sences, or internall operations of the soule." Physical, not spiritual qualities, were becoming the principal object of regard to a significant number of students of animals and man. Petty is merely illustrative of this tendency. He and the many others who approached the primates from the physical point of view emphasized the great classificatory and anthropological importance of the shape of the nose, lips, and cheekbone, the facial outline and form of the cranium, the texture of the hair and color of the skin. By 1735 this practice was well advanced. In that year appeared the first edition of Carolus Linnaeus' *Systema naturae,* the critical work which, especially in its later editions, led to a wholesale recasting of animal classification and laid the foundations for a secure and coherent science of natural history. Linnaeus not only included man in his catalogue of the living world (designating him *Homo sapiens*) but attempted, on the confused grounds of physical and cultural features, to draw sharp distinctions among the varieties of mankind.

By the nineteenth century, consequently, the scrutiny and tabulation of man's physical characteristics were well underway. The dramatic success of comparative anatomy at the century's turn, a combined effort of French, German, and British workers, served only to accentuate the realization of man's animal features. Anatomy and ethnology worked hand-in-hand throughout the nineteenth century in common pursuit of effective criteria for the discrimination of human racial varieties. Amid this welter of divergent physical features it was evident that, in the present world at least, the accustomed unity of mankind would be severely tested. Seemingly permanent races of men were described, their definition was emphasized by association with long-familiar prejudices and, despite its heterodox implications, the notion of aboriginal differences among men returned to favor. Our anthropologists were modern Western men and their increasing contact with strange peoples now supported less the pleasurable flavor of exotic appearances and customs than an earnest effort to put order into the confusion of the human condition. The easy and accepted solution was to isolate the superior races and determine the varying degrees of inferiority of other peoples.

Ethnocentrism was nothing new to the nineteenth-century Western world. Aristotle had separated Hellene from Barbarian. The latter, victim of inferior circumstances, was yet a savage. In that condition he was no better than a beast, for he lacked civilization, the truest mark of the human condition and evidenced by law, religion, security, and sexual continence. Man too easily views other men through the limited spectacles of his own devising. The universal claims of Christianity in rigorous form reached out to all mankind but extended access to salvation not to all men as men but only as Christians. The Christian world was poorly prepared for a dispassionate assessment of human bodily and, above all, cultural differences. The eminent moral philosopher, David Hume, was wonderfully explicit on the matter of human unity. "I am apt to suspect the negroes," he wrote in 1748,

"and in general all the other species of men (for there are four or five different kinds) to be naturally inferior to the whites. There was never a civilized nation of any other complexion than white, nor even any individual eminent either in action or speculation. No ingenious manufactures amongst them, no arts, no sciences. . . . Such a uniform and constant difference could not happen, in so many countries and ages, if nature had not made an original distinction betwixt these breeds of men.

Hume's statement displays how easily a vast complex of human traits and behavior—"civilization," with all of the cultural details that that term denotes, mental capacity and physical appearance—were confused. Of critical importance, of course, was the implication that skin color was a reliable indicator of other or all human qualities.

The confusion of anthropological interests and racial prejudice probably reached its peak in the so-called American school of anthropology of the immediate pre-Civil War period. This school, including Samuel Morton, a Philadelphia anatomist, George Gliddon, an English-born popularizer of the sciences, and Josiah Nott, a physician from Mobile, Alabama, were particularly concerned with physical anthropology and were all, despite the public piety of the times, pugnacious polygenists. Their study of man was an uneven mixture of scientific interest and social commitment. Their ostensible scientific proofs (measurement of cranial capacity) were joined with the suggestions of archeology and written histories to demonstrate that mankind evidenced, in its earliest ascertainable condition, just those deep-seated racial differences so conspicuous and necessary today. The Negro's ancient inferiority was striking confirmation that he was today no degraded being but enjoyed his inferior status as part of the very constitution of our world. He was,

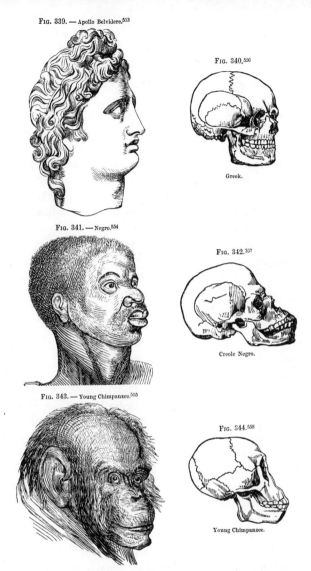

FIG. 339. — Apollo Belvidere.[553]

FIG. 340.[556]

Greek.

FIG. 341. — Negro.[554]

FIG. 342.[357]

Creole Negro.

FIG. 343. — Young Chimpanzee.[555]

FIG. 344.[558]

Young Chimpanzee.

Figure 5.1   Physical anthropology has served many masters but none, perhaps, exacted so explicit a message as did slavery apologetics in antebellum America. Josiah Nott verbally and visually urged, on ostensibly scientific grounds, the near-bestial condition of a distinct Negro "race" of mankind: "Although I do not believe in the intellectual equality of races, and can find no ground in natural or in human history for such popular credence, I belong not to those who are disposed to degrade any type of humanity to the level of the brute-creation. Nevertheless, a man must be blind not to be struck by similitudes between some of the lower races of mankind, viewed as connecting links in the animal kingdom; nor can it be rationally affirmed, that the Orang-Outan and Chimpanzee are more widely separated from certain African and Oceanic Negroes than are the latter from the Teutonic or Palasgic types." (Josiah Nott and George Gliddon, 1854.)

of course, a most suitable person for menial tasks and the condition of slavery.

This association of the seemingly serious study of man and racist sentiments was a prolonged phenomenon. While the literature of the subject is vast and modern studies numerous, they must not exclude from this discussion at least brief notice of other active areas of nine-teenth-century concern with man's place in nature. Perhaps only a minority of the physical anthropologists, who were largely polygenists, were motivated by explicitly racist intentions. An area of great interest to these men and one no less a great trial to literal interpreters of *Genesis* was human paleontology. Prior to 1850 no irreproachable fossil human specimen was known. By irreproachable is meant the demon-strable association of the specimen with the indubitable antiquity of the mineral deposit in which it was found. In 1856, however, the first generally certified fossil human remains were unearthed in the Neander Valley of Germany. The suggestive find of Neanderthal man was soon joined by the discovery (1868), in southwestern France, of Cro-Magnon man (clearly a variety of *Homo sapiens* and appearing in Europe about 35,000 years ago). Ernst Haeckel had long insisted that a truly complete evolutionary schema demanded a "missing link" between the anthropoid apes (gorilla, chimpanzee, orangutan, gibbon) and the hominids or man. Inspired by Haeckel's plea, Eugène Dubois, a Dutch paleontologist, searched and found (1891) on the island of Java the remains of the most celebrated of all prehistoric men, *Pithecanthropos*. And in the twentieth century a still more ancient form, *Australopithecus*, has been uncovered in East Africa. These specimens all date from the Pleistocene epoch, roughly the last million years, and thus confer on man an unanticipated antiquity.

Human paleontology, powerfully suggestive however incomplete be its evidence, thus joined comparative anatomy in binding man ever more closely to his animal antecedents. Huxley, an able student of paleontol-ogy as well as of anatomy, keenly sensed the importance of these inquiries.

"Brought face to face with these blurred copies of himself [the reference is to the anthropoid apes], the least thoughtful of men is conscious of a certain shock, due perhaps, not so much to disgust at the aspect of what looks like an insulting caricature, as to the awakening of a sudden and profound mistrust of time-honored theories and strongly rooted prejudices regarding his own position in nature, and his relations to the underworld of life."

Certainly no evidence can be so persuasive of man's antiquity and of his roots in the animal kingdom than that returned by paleontology, presenting as it does concrete, albeit disconnected, instances of the actual course of human transformations. Despite the often ferocious disputes that surrounded the identification and interpretation of certain of the fossil forms, recovery of the physical traces of man's past probably stands highest among the nineteenth-century biologists's contribution to our understanding of man's place in nature.

A second domain of exceptional activity and great achievement regarding man's past was prehistoric archeology. Interest in human antiquities passed during the nineteenth century from the collecting of pleasant or curious artifacts to the systematic search for and interpretation of these traces of extinct human cultures. Flint tools and weapons, fragments of pottery, worked pieces of bone, and countless other items provided the raw materials for the archeologist. The science progressed slowly and was duly cognizant of scripturally imposed limits on the possible antiquity of mankind. Broad classificatory divisions were nevertheless devised; the Dane, Christian Thomsen, created the familiar categories of Stone, Bronze, and Iron Ages in 1836. And the accumulation of reliable materials continued, the exploration of cave-deposits of the late Pleistocene epoch being particularly valuable.

Perhaps the decisive work in the creation of archeological science was the activity of a French provincial customs official, Jacques Boucher de Perthes (1788–1868). Canal dredging in the Somme Valley of northern France disclosed shaped stones which Boucher de Perthes concluded were ancient tools (hatchets). They were found in close association with bones of (extinct) animals. In *Antiquités celtiques et antédiluviennes*[1] (1847–1864) he argued that this evidence clearly showed the existence and activity of prehistoric men. Similar finds were made in English caves during the 1850s. But recognition by the scientific world at large of the exceptional importance of this work only came in 1859 when an official delegation of the Royal Society of London announced its conversion. The archeologists clearly had shown the existence of Pleistocene man.

The subsequent and great advances of archeological investigation regrettably cannot form a part of this brief history of biology. The coincidence of events toward 1860 demands, however, special emphasis. Several trains of argument and evidence were converging. Archeology, human paleontology, and the Darwinian transformation hypothesis all

---

[1] *Celtic and antediluvian antiquities.*

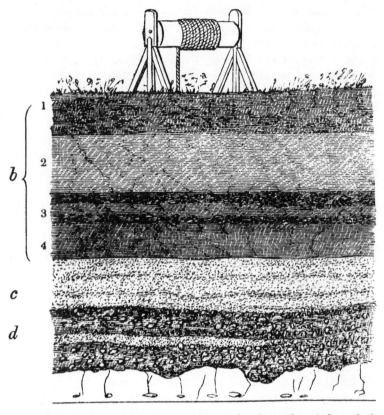

Figure 5.2   The remarkable discovery by Boucher de Perthes of the undeni-able presence in undisturbed alluvial deposits of both flint tools and the remains of extinct mammals was confirmed by an official delegation of the Royal Society of London. The report to the Society included this sectional drawing of a pit at St. Acheul in the Somme Valley. The "d" level, five feet below the surface, contained the sought-after association of flints and fossils. Levels "a" and "b" offered no organic remains and the presence of flint implements in level "c" was questionable; below all this material, and there-fore oldest of the beds examined, was the level that demonstrated the coexistence of ancient man and extinct animals. (Joseph Prestwich, 1860.)

joined at this time and they appeared to express a common message: man and human society were demonstrably ancient; human prehistory might well be recreated; man was an animal and as such probably was subject to the same transforming forces as other creatures. Arguments of tremendous bearing were thus being prepared for those who chose

not only to place man in nature but to abstract from him all super-
naturally provided spiritual qualities. Here were adequate grounds for
compelling a fresh assessment of the nature of man and the meaning
of his history.

A Social Animal and His History

Man, Aristotle had argued, was of necessity a social animal. Divorced
from the company of other men, he failed to realize his singular powers,
remained unhappy, and was lacking in essential virtue. It was just these
qualities that defined this creature, man, and in their absence—a consequence of an isolated existence—true humanity was unattainable. This
was a perennial argument and one that has informed most subsequent
Western consideration, be it Christian or strictly secular, of the organization of human affairs. Whether or not the social condition is a uniquely
human manifestation is of less interest than the fact that society is a
distinctive feature of mankind. As such it may offer terms for closer
and special analysis and suggests, furthermore, the possibility of erecting
an autonomous science of society. This option was boldly seized by the
French sociologist, Emile Durkheim, in the closing decades of the
nineteenth century. His passionate concern with man's social condition
had been anticipated, however, by several generations of observers,
recruited principally from history and moral and political philosophy.

As historical explanation, the acute historical sense characteristic of
Western civilization has already been recognized in this volume. In the
200 years following about 1650, there matured first a recognition of
the superiority of modern over ancient thought and action and then the
firm conviction that this progressive change might be truly open-ended.
In its most optimistic form, attained towards the end of the eighteenth
century, the idea of progress had freed itself from all real constraint.
Man in the past had evolved from a socially formless hunter stage into
a pastoral nomad and had then reached a settled and socially stable
agricultural stage. His ascent next brought him to industrial activity and,
ultimately, to the unheralded complexities of urban society. Moral and
intellectual development accompanied, and provided the critical dynamic, for the social evolutionary process. Naturally, only Western man
had yet attained these highest levels.

These themes formed the substance of what came to be called at
times universal, at other times conjectural history. A leading spokesman
of the art, the French statesman, Anne Robert Jacques Turgot, de-
claimed in 1751 that the

"continual combination of his progress with the passions and events which these have produced constitutes the *history of mankind*. Here each individual is no more than a part in an immense whole which, like man himself, presents its childhood and development. *Universal history* therefore embraces the consideration of the successive changes of mankind and the study of the causes which have brought them forth."

Forty years later (1794) the Marquis de Condorcet reiterated and developed this grand doctrine, dramatically pointing up moral progress in commenting that "we pass by imperceptible gradations from the brute to the savage and from the savage to Euler and Newton." From Condorcet's classic pronouncement (*Esquisse d'un tableau historique des progrès de l'esprit humain,* 1795),[2] from new German metaphysical and historical doctrines, and from the writings of the Scottish moral philosophers, the nineteenth century received profitable instruction in the requirements of universal history.

Universal history deemphasized the individual. Moreover, exact chronology was less important than recognition of major historical movements such as the nature and sequence of the stages of man's evolution. It is supremely important that universal history provided a legitimate place for undated man, that is, man who had yet to rise to a truly civilized condition. The primitive condition thus became a genuine and essential component in the total human evolutionary process. By 1800 the great chain of being had become more a temporal process than a fixed plan for the disposition of the presently existing order of beings, including plants, animals, and men. From the viewpoint of the study of man the universality of this process not only assured students that primitive man enjoyed a necessary (and, of course, inferior) place in the explanatory scheme but also recalled the fundamental scriptural claim for the oneness of mankind.

During the nineteenth century, universal history, doubtlessly far more than scriptural interpretation, environmentalism, or even transformist doctrines, offered a ready solution to the patent and disturbing fact of human diversity. And with this solution emerged the century's predominant anthropological investigatory weapon, the comparative method. Unity of mankind—this need and "fact" universal history readily explained by tacitly or explicitly accepting the orthodox account of the creation and postulating a common developmental sequence for all men of all times and nations. This latter fact or, rather, extravagant

[2] *Sketch for an historical view of the progress of the human mind.*

hypothesis then resolved the question of human diversity. Men were different because they had attained to differing levels on a unique, progressive evolutionary scale. All men possessed the potential to advance; some had done so and some had not; future changes were certainly to be expected.

A French Jesuit missionary to French Canada, Joseph Lafitau, is commonly credited with the earliest declaration and full exploitation of the comparative method. Lafitau had hoped to comprehend the mores of Greco-Roman antiquity, and in the Canadian wilderness he discovered the living solution to his quest. "I have been satisfied," he remarked in 1724,

"to understand the character of the savages, and to make myself acquainted with their customs and practices. I have searched among these customs and these practices for traces of the most distant antiquity; I have read with care those of the most ancient writers who have treated of the manners, laws and usages of the peoples with whom they had some acquaintance; I have compared these manners with one another, and I confess that while the ancient writers have given me lights on which to base some lucky guesses concerning the savages, the customs of the savages have given me light to understand more easily, and to explain, many things which are in the ancient authors."

The comparative method was a remarkably simple device. Granted a common course for all human development one might seize any particular slice of time—for example, the present—and expect to find among the great existing heterogeneity of peoples scattered about the world materials for the reconstruction of the totality of human history. Above all, modern savages (Lafitau's Canadians) could be used to help recreate the manners and thought of nations whose remains had long since been obliterated. The determinative premise throughout was the real comparability of seemingly divergent human societies. The practice, like the biologists' recapitulation doctrine of which the comparative method is the probable antecedent and definite conceptual ally, was open to scandalous abuses. Lafitau used Eskimo canoes and boats, for example, to interpret ancient Egyptian inscriptions. Such efforts, some more extreme and others more cautious, were repeated *ad infinitum* throughout the nineteenth century and doubtlessly contributed to making the comparative method the most disreputable of procedures to the modern anthropologist.

It was in the context of these comparative studies that anthropology

Figures 5.3 and 5.4   The analysis of familial relationship, itself an excellent clue to the fundamental structures of a society, is a primary objective of modern anthropology. A valuable tool for such analysis is the accumulation and tabulation of the terms current in a society for designating these relationships. Morgan's famous diagrams of the English system (5.3) and Seneca-Iroquois system (5.4) of consanguinity, together with the vast mass of supporting data from which the latter were derived, provide an important historical basepoint for the linguistic approach to kinship analysis. (Lewis Henry Morgan, 1871.)

defined its early goals and constituted itself as a science. This achievement was largely the work of the so-called evolutionary anthropologists of the post-Darwinian period. The leading evolutionary anthropologists, including the English prehistorian John Lubbock (1834–1913), the American lawyer Lewis Henry Morgan (1818–1881), and Edward Tylor (1832–1917), a masterful and critical exponent of the evolu-

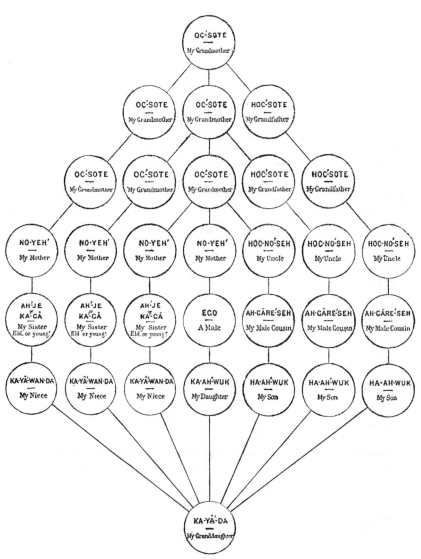

tionary viewpoint and the study of culture, argued that man as a social being creates human culture. Culture, an omnibus term that embraced mythology, kinship organization and terminology, technology and the arts, language, and morality, was consequently an historical product. Man in the savage condition exhibited minimal cultural attainment;

man in nineteenth-century industrial Europe represented the highest human attainment of civilization. Strict continuity of the evolutionary or historical process bound together savage and European and all their intermediaries. While allowing for occasional deviations from the advancing course of human history and emphasizing "survivals" of ancient cultural artifacts in modern societies the evolutionary anthropologists were generally advocates of progress and the validity and usefulness of the comparative method.

Already, however, the question of the uses of anthropology had arisen. The comparative method might illuminate either past or present. The objective of one group of investigators, including Lafitau and Morgan (the latter's investigations of Iroquois and Ojibway language usages (1871) opened into the vast discipline of kinship analysis), was to recover the past. One did so by assaying the cultural wealth and variety of modern savages and lesser cultural stages and arranging these in (hypothetical) historical order. But others, notably Tylor, sought better to understand his own, that is, advanced cultural condition. Only historical explanation could provide such understanding. He stated in 1865,

"It is indeed hardly too much to say that Civilization, being a process of long and complex growth, can only be thoroughly understood when studied through its entire range; that the past is continually needed to explain the present, and the whole to explain the part."

Tylor did not, therefore, engage in large-scale historical reconstruction. He sought instead to seize and assess the common and persistent elements of human cultural experience. Morgan's efforts, however, led to a major monument of conjectural history, *Ancient society* (1877). Herein the stages of human progress were intimately related to the material conditions and technological refinement of society. Morgan's argument greatly impressed Friedrich Engels and his book remains in high favor among Marxian dialectical materialists.

These uses of the comparative method, and the general orientation they imposed on late nineteenth-century anthropological thought, probably represent divergent tendencies rather than a rigid dichotomy. Whatever his objective was, the essential premise of the evolutionary anthropologist was the historical nature of man and culture, of society and its products. As one examined primitive or uncivilized man one's attention halted less on the individual and more on the composition of the society of which he formed an essential but less interesting part. This fact seems an inevitable consequence of the comparative method and of the aspiration toward universal history in whose context that method found both inspiration and justification. The nineteenth century announced itself an age of individualism, yet its scholars and critics increasingly pursued and vaunted the social aspect of man. In the 1890s this appeal for a science of society, a sociology, passed from aspiration to preliminary formulation.

Students of Society

In the doctrines of Auguste Comte may be found the principal stimulus for self-conscious social analysis. Comte (1798–1857), a French social philosopher fanatically devoted to the twin ideals of social order and progress and elaborate intellectual systematization, had carefully specified both a unilinear evolutionary sequence for society and a classification of the sciences which assured sociology (the term is Comte's) high status. His rule of the three stages of thought and society—theological, metaphysical, and positive (or scientific)—announced the inescapable and progressive historical adventure of the human mind. Comte was the inheritor of the eighteenth century's dogmatic elevation of reason to supreme position among human qualities. In the positive stage reason finally stood free of long-standing and harmful emotional and doctrinal tendencies or commitments. The positivist was scientific, meaning that his ideas were clear and secure, being ever based on definite phenomenal references. The essence of things he did not seek; his quest was for exact specification of the relationship between two or more events. He favored the mathematical notion of function, hoping always to specify in the sciences of nature or of society the exact terms denoting how change in a fundamental phenomenon is related (or causes) change in a dependent phenomenon.

Comte announced (most systematically in *Cours de philosophie positive,* 1830–1842)[3] that the sciences themselves presented a hierarchy. Mathematics, the science of pure relation, had long been positive, and astronomy, physics, and chemistry had, in that order, also become positive. In the past generation biology, too, had won its scientific spurs. The time was ripe for the elaboration of the penultimate science, sociology (ethics, as the foundation of Comte's later and unusual Religion of Humanity, was presumably the ultimate science). The effect of this schema, whatever were the merits of Comte's highly personal

[3] *Lectures on positive philosophy.*

speculations on the nature of human society, was to place the study of society inextricably among the sciences. Here was a proclamation of faith and a program to be exploited by future students of society. Comte's objective was a familiar article to later nineteenth-century students of society. Yet, despite the great prosperity of positivist doctrine and the vocal if quaint body of Comte's disciples, his profound hopes for the formulation of a useful method and sharper definition of a genuine and autonomous science of society found first realization in Durkheim's publications of the 1890s. But Comte's vision first came to haunt the self-regard of Herbert Spencer, a sometime railroad engineer turned political philosopher and metaphysician. Spencer during the 1850s and more fully in his treatises on sociology published between 1873 and 1896 had also cast foundations for the erection of a new science of society. While sharing with Comte a general evolutionary bias and meticulous concern for classifying the sciences, Spencer felt that his system was born wholly free from any influence by the French philosopher's doctrines. Spencer may have been generally correct but throughout his career he manifested acute sensitivity to charges that his views derived from his predecessor (Comte himself had angrily denied any debt to his master, Henri de Saint-Simon).

Spencer (1820–1903) was the most popular philosopher and prolific metaphysician of the Victorian period. His idiosyncracies were as singular as the relentless single-mindedness of his grandiose philosophical system, styled the Synthetic Philosophy. Like Comte's Positive Philosophy Spencer's system was highly deductive and designed to embrace the totality of human experience. Again like Comte, Spencer sought to place sociology rightly and firmly among the sciences. But the similarities between these two doctrines, customarily regarded as the ideological base-points for the subsequent definition of the scope and methods of the social sciences, must not obscure the fundamental differences that separate them. Comte and Spencer were equally evolutionists, paying great heed to the historical transformations of human society, but their specific evolutionary doctrines were radically unlike one another. Comte advocated a rigid, unbranching scale of ascent while Spencer, also progressive in spirit, emphasized the limitless diversification of the evolutionary products. Their doctrines suggested, of course, quite different conclusions. They differed no less deeply over the precise place to be accorded sociology among the sciences and how the new discipline might draw upon its lesser scientific partners.

Although founded on the logical order of the increasingly positive sciences, sociology, Comte had argued, must necessarily define its own interests and means of inquiry. All sciences were ultimately defined by their subject matter, that is, the phenomena of which they treat, and for sociology that subject must clearly be human society. One did not, consequently, truly derive or base sociology on biology, among whose interests was, of course, man as an animal. Spencer would have none of this. To him the sciences formed a unity. The content of biology was critically important to any consideration of society. Comte's narrow concentration on reason or ideas as the dynamic of the evolutionary process, exemplified by the rule of three stages, wrongly eliminated what Spencer held to be the ultimate elements (matter and motion) in any valid metaphysical system. Spencer's total system was, philosophically speaking, realistic, not idealistic like those of Comte or, later, Durkheim.

Spencer integrated biology fully into his philosophical system and prescribed preparation in this science (and also in psychology) as the necessary preliminary to approaching social questions. The importance of biology is that it provides "an adequate theory of the social unit—Man. For while Biology," Spencer continued (1873),

"is mediately connected with Sociology by a certain parallelism between the groups of phenomena they deal with, it is immediately connected with Sociology by having within its limits this creature whose properties originate social evolution. The human being is at once the terminal problem of Biology and the initial factor of Sociology."

The "properties" of man "originate social evolution"—this expression implicates Spencer's conception of individual man, social organization, and the evolutionary process, thus, obviously, the importance of biology. Man and society take their place in the cosmic evolutionary process. This process, Spencer announced, moved inexorably from the homogeneous to the heterogeneous, that is, from an amorphous, utterly undifferentiated ground substance (matter) to formed and progressively more complex entities. The essence of all evolutionary change —and such change was the paramount feature of cosmic design—was relentless and progressive differentiation. The process began in the inorganic world, advanced to the lower forms of life, and moved on to the higher animals and man. That same evolutionary process continued to direct the social organism. Spencer, no less than Comte and most nineteenth-century social theorists, freely invoked the ancient metaphor of the social organism. The metaphor announced that society, like the plant or animal, was a dynamic totality whose functional integrity, whatever its origin, was maintained by the simple fact that all parts and processes were neatly adapted to one another and to the creature's place in nature. In the normal or healthy condition, the organism—be it plant, animal, or social—thus preserved its individuality and prospered according to the terms set by its form and functions.

Spencer's evolutionary social philosophy later acquired a gloss of Darwinism and came to appear coeval with that doctrine. But it had been sketched out during the 1850s, quite in ignorance of Darwin's maturing and still unpublished hypotheses, and preserved the independence of its central concern, social analysis as the basis for social action or abstinence from action. Behind all metaphysics and science Spencer's target was not unlike that of Comte and, later, Durkheim. One did not in consummate dispassion simply dissect society; one studied society in order to lay intelligent and effective grounds for coping with the daily affairs of human intercourse, the broad domain of politics. Comte in his Religion of Humanity offered a distinct notion of the needs of an ideal society. Spencer drew upon his own analysis of society in order to portray the dangers inherent in making indiscriminate or sudden large-scale changes in society. Spencer's conception of social change was evolutionary, not revolutionary, and formed part and parcel of a conservative political philosophy.

It was the individual that dominated the thought of Spencer, the student of social phenomena. In political terms this entailed defense of prevailing *laissez-faire* economic apologetics and appreciation of the "steadying effect, alike on thought and action" which evolution doctrine would work on too ardent social reformers. In terms of social analysis it suggests the really traditional quality of Spencer's sociological ambitions. Society he regarded not as an object *sui generis,* exhibiting unique properties and demanding new and peculiar tools for its study. Society was instead an aggregate, a virtually mechanical aggregate of the many elements that entered into its formation. These elements were individuals—atoms and molecules for inorganic aggregates, cells for plants and animals and individual men for human society. "In every community," Spencer maintained,

"there is a group of phenomena growing naturally out of the phenomena presented by its members—a set of properties in the units. . . . Setting out, then, with the general principle, that the properties of the units determine the properties of the aggregate, we conclude that there must be a Social Science expressing the relations between the two."

Just as the spherical mass of the cannonball will limit the shapes of the piles into which these balls may be formed, or the quality of the brick determines the height and strength of the wall, so will the social unit, man, present the ultimate grounds for casting a science of society. For Spencer, human society was neither more nor truly other than the components of which it was formed. Its peculiarity resided in the complexity of its aggregation and its interest derived from its irreducible element, man as an individual.

In Spencer' system the cohesiveness or solidarity of society was more apparent than real and clearly was no primary desideratum. For Durkheim social solidarity was at once an essential fact and the touchstone for all valid social analysis and action. Emile Durkheim (1858–1917) had been trained in philosophy but after 1880 dedicated himself to the creation and furtherance of an autonomous science of society. That dedication brought great reward; through his pupils, personal research, and publications on sociological method, Durkheim by 1900 had emerged as the foremost figure in his discipline. His doctrines were already reaching beyond his native France and were disclosing their relevance not only for analyzing "higher" or industrial Western social forms but for examining the structure of "primitive" or preindustrial and non-Western peoples. Durkheim's views underlie both modern sociology and social anthropology.

Implicitly pursuing Comte's argument, Durkheim urged that "social facts" have a reality of their own. These facts—moral and legal demands, religious doctrines, financial and linguistic systems, and numerous others—exist outside and independent of the individual. They are common to society at large, are conveyed to the individual through example, admonition, and formal education, and act at all times as a constraint upon the individual. The social fact is the phenomenon of interest to the sociologist and the phenomenon must determine the objective and procedures of the science.

Durkheim claimed to refuse all commerce with what he called "ideology" in the social sciences. Comte's crime was the rigid structuring of social analysis upon an arbitrary evolutionary scale, topped off by a hopeful expectation of the coming reign of Humanity. Spencer's sin was mindless individualism, a commitment which, in Durkheim's opinion, precluded any possibility of an autonomous social science. Comte and Spencer shared, furthermore, a common vice: they offered historical explanation instead of causal analysis. This charge touches the heart of Durkheim's epistomological position and with it his view of the prospects and responsibilities of social science.

Scientists, Durkheim revealingly declared, have assumed as part of their method the logic of causal connection. Only philosophers, and there is a note of scorn in the claim, have seen fit to question causality. To Durkheim causality expressed only a constant conjunction or association. Causal science, as Comte in his more positivistic mood had announced, is but a science of relations. It has nothing to do with chronological sequence. The stages of history merely follow one another; they do not "engender one another." The scientist who turns to history is no scientist, for, grasping but an orderly arrangement in time, he will find no dynamic for the process. Like the physiologist Claude Bernard, whose views may well have influenced his study of the social organism, Durkheim shunned seeking the essence of being or process in order securely to correlate under rigorously controlled conditions the phenomena, our only certain knowledge, of social structure and behavior. In this sense Durkheim no less than Bernard and countless other late nineteenth-century thinkers was a doctrinaire positivist.

His conclusions were indeed revolutionary. Durkheim deliberately rejected the appeal to history that had for generations girded the explorations and conclusions of Western social thought. Historical explanation was evidently no explanation at all. It was philosophically invalid and must be unsatisfying to self-critical minds. Comte's perpetuation of monolithic and progressive historical schemes and Spencer's articulation, based on the flexibility of the individual, of continually diversifying evolutionary process stood only as signs of outmoded preoccupations. The thrust of Durkheim's thought was radically atemporal or presentist and this ahistorical bias has left an indelible mark on subsequent generations of social scientists.

No less profoundly did Durkheim disjoin biology and sociology. Spencer's demand for preparation in biology simply had no meaning. Spencer had argued that

"there can be no understanding of social actions without some knowledge of human nature; there can be no deep knowledge of human nature without some knowledge of the laws of Mind; there can be no adequate knowledge of the laws of Mind without knowledge of the laws of Life. And that knowledge of the Laws of Life, as exhibited in Man, may be properly grasped, attention must be given to the laws of Life in general."

Not only was Durkheim unconcerned with the laws of life in general, he announced that the laws of mind or (individual) psychology were simply irrelevant to social analysis. From the Scottish moralists and

French social philosophers of the Enlightenment through Comte and Spencer, critical terms for social analysis had been grounded on strictly psychological commitments, most notably the psychic unity of mankind. The nineteenth-century social philosophers infused this idea with suggestions from biology which bore particularly on the probable physical bases of mind in the brain and nervous system. It is through these developments that Spencer's argument gains its force. But Durkheim and his supporters were asserting the autonomy of the social sciences. There were accessible facts that were demonstrably social in nature and in need therefore of a new science. That science was thus designed to cope with these facts; it had no need to draw on doctrines or methods designed for and therefore uniquely suited to either psychology or biology.

Durkheim's theoretical stance was fully stated in the incisive *Règles de la méthode sociologique* (1895).[4] He was, however, much more than a theoretician. He carried out a number of arduous and exemplary social analyses, most notably on the moral and legal codes that bind the members of society into a cohesive body and on the incidence and social meaning of various kinds of suicide. His work demonstrated the power of fresh instruments for social analysis (statistics) and was rich in new concepts (particularly the idea of anomie, a state of crisis in the social order brought on by a weakening or confusion of the regular norms of the society). In his later career Durkheim turned to studying the "collective consciousness" of society and the forms of religious practice and belief.

Through his own influence and that of his disciples Durkheim's mark was soon set upon sociology. Not least did it lead to intimate personal involvement of the inquiring student with the social phenomena under inspection. Sociologists began to harvest and assess data drawn directly, through participation, observation, and questionnaire, from society. Anthropologists, loosely defined as students of man in his primitive condition, had long relied for their information on printed accounts and imperfect reports of travellers. By century's turn they, too, had commenced serious field work. Durkheim's influence was widely felt in anthropology only in the 1920s. This came about through the advocacy and example of the great modern triumverate of Marcel Mauss, Durkheim's nephew and an able anthropologist as well as sociologist, Alfred Radcliffe-Brown, theoretician and progenitor of the British

---

[4] *Rules of sociological method.*

school of social anthropology, and Bronislaw Malinowski, whose famous *Argonauts of the Western Pacific* (1922) dramatized the ideal and established high standards for systematic field work in anthropology.

Man

The most salient feature of the rapidly advancing nineteenth-century studies regarding man was their general presumption that he, too, was or must soon become a proper object of scientific investigation. Science was reaching out to seize the supreme product of creation. With science came the assurance that its admired qualities—certain knowledge grounded always in facts, experimental control over phenomena, and rigorous logical systematization, the whole giving promise of predictive power—might extend from the inorganic realm to plants and animals and, ultimately, to man and human society.

But this obvious and commonplace conclusion obscures the probably more fundamental diversification of the sciences of man that was also taking place. Only license permits one to speak even loosely of a "science of man." In reality, manifold aspects of man were being carefully recorded. Depending on one's point of view and predominant interest, one selected among these aspects those whose emphasis the desired analysis required. In the Christian tradition man presented, of course, a most ambiguous face for scrutiny. His essential being was spiritual and he was properly the exclusive charge of moral science. Renewed attention during the seventeenth and eighteenth centuries to his physical qualities and apparent historical antecedents could not fail to emphasize his relationship to other members of animate creation. The portrayal of man as an animal joined other critical, nonbiological tendencies in Western thought to produce during the nineteenth century a serious and widespread crisis of confidence in the still prevalent Christian conception of human nature, history, and society. Concrete manifestations of the new spirit may be witnessed in the definition during this period of the sciences of physical anthropology, human paleontology, and prehistoric archeology.

Another perspective on man was social. The distinctive quality of man as a social being found emphasis in Greek thought and continued to exert an influence on all serious Western reflections upon the nature of man. To men of long experience the continuity of human society was not only a fact but also came to assume a causal bearing. History had meaning both as a curious sequence of peoples and civilizations and as the vital dynamic that ensured that generations did not simply follow but truly produced one another. Historical explanation offered the conceptual grounds for universal history and for elaboration and implementation of the comparative method. With great (Spencer) or with minor (Comte) allegiance to biology and its apparent dictates, the nineteenth century assembled the scattered strands and pronounced the possibility and necessity of creating a genuine science of society. A sense that this high ambition was not vain was given by Durkheim. His endeavor redirected scientific interests to current societies and largely purged sociology and anthropology of historical content. His methodological prescriptions made investigation of social facts truly practicable and invited exploitation. The twentieth century testifies to the triumph of the social sciences.

CHAPTER VI

Function: The Animal Machine

PHYSIOLOGY UNDERWENT EXTRAORDINARY DIVERSIFICATION during the nineteenth century. Its subject matter expanded and its objectives were transformed. To the traditional investigation of the function of organs and organ systems were joined increasingly fruitful analysis of cellular processes and numerous invitations to reduce all such events to the terms of seemingly more fundamental sciences, notably physics and chemistry. On each of these levels the physiologist could investigate and, hopefully, explain the multitudinous and notoriously complex vital phenomena whose summary effect was, for simplicity's sake, called life. The physiologist might explore, for example, the action of the liver. Among its functions was discovered the capacity to produce animal starch (glycogen). The investigator could halt here and rightly announce this fact as an exceedingly important organ-level statement of physiological activity. This conclusion would not and did not, however, satisfy all physiologists. Demanding more specific knowledge of the process of glycogen formation these men sought further to localize that process and, by extension, all comparable physiological processes within the organ. With the announcement of the cell theory their ambitions gained a conceptual foundation upon which to build; physiology by 1870 was much occupied with cellular processes and how these might be harmoniously integrated to produce the complete organism. But a yet further level of resolution seemed already at hand, that of regular physico-chemical or mechanistic processes that might be presumed to offer a fundamental, even irreducible basis for both cellular and organ-level phenomena. To the more impassioned advocates of this view, the uniqueness of vital processes was apparent only and was due largely to the simple ignorance of physiologists. Happily, those of the mechanistic faith believed the end to such ignorance was near; unhappily, their hopes long sounded more loudly than their concrete accomplishments.

Amid the welter of the century's physiological interests one complex of problems compels special consideration. Broadly stated, this complex includes the processes of respiration and important aspects of digestion and excretion. Respiration serves the paramount function of providing the energy that powers the living being. It involves a transfer and production of gases and, most importantly, the liberation of heat and energy in a form useful to the organism. The steady production of animal heat had been the object of speculation and study since classical antiquity. That "air" (a substance believed, until the mid-eighteenth century, to be homogeneous in nature) was in some mysterious way indispensable to life was also an ancient conviction. During the nineteenth century, however, the overall chemical and physical relationships of the respiratory processes—which subsume the question of organic heat production—were brought into close agreement with the principle of the conservation of energy. The latter generalization was itself a major contribution of the era. These events demonstrated that the living creature, whatever else it might be or whatever it might be claimed to be, was an integral part of the physical universe. Its very being and sensible reactions—movement, electrical and chemical phenomena, and perhaps even conscious behavior—all depended on the availability of energy. The investigations that founded this new viewpoint constitute one of the greatest achievements of physiological science and offer the theme of the early sections of this chapter.

The conclusion that the organism was a veritable heat engine obviously had required research based on physical and chemical methods. Many physiologists believed that the success of this and similar research proved what men from many callings had long maintained, that the organism was nothing but what prevailing explanatory concepts of the physical sciences might require. Thus life was deemed the mere product of matter and motion or of the action of the essential forces animating the cosmos, the opinion maintained by radical materialist, mechanist, or reductionist (these terms are not synonymous but their passing definition causes more difficulties than it resolves). Others denied that any valid statement regarding the essence of life, of the very nature of the organism, and the meaning of its functional affairs, could be extracted from the simple fact that vital processes could be explored, described, and discussed in physical or chemical terms. Still others claimed that the organism, by its spontaneous behavior and dismaying obstinacy before experimental interrogation, betrayed the presence of forces unlike those of the physical realm, forces indeed peculiar to living things. At every stage in the progress of physiology during the nineteenth century will intrude these alternative interpretations of what life and vital processes may truly be. They raise important questions regarding physiological method, if such there be, and bring one close to related developments in philosophy and popular thought. The physiological interpretations and these questions will be examined in the later sections of this chapter.

The Animal Machine

By the late seventeenth century two distinctive yet interdependent meanings of machine had entered English usage. On the one hand a machine was a contrivance, however simple or complex, for the application of power to a particular task. On the other hand machine referred to a combination of interrelated parts that performed their destined operations mechanically, that is, without occasional intervention or sustained regulation by voluntary (conscious) or unconscious action. With regard to the living organism, machine in the first sense was meaningful strictly within limits imposed by the many ambiguities of the term power. It is from this realm of definition that the issue of irreducible vital forces may best be viewed and the ultimate equation of energy (as defined by the physicist) and vital force(s) considered. Intrinsic to the second definition is, of course, a disclaimer of action on the psychological or, read more generally, biological level. In its radical form, it was this meaning of machine that revivified the perennial debate over the existence of a true animal machine. On the outcome of that debate and also as part of its initial postulates revolved such ultimate questions as the relationship between animal and man and the very definition of man himself.

Descartes, whose writings (*Traité de l'homme,* 1664; *Passions de l'âme,* 1649)[1] opened the debate for the modern era, maintained that an animal was an automaton, lacking both sensation and self-awareness. Man, however, was endowed with a soul which mediated both of these functions and also bound him, as Christian tradition required, to his creator. Elucidating the precise nature of the relationship between body, or machine, and soul offered challenging substance to many a Cartesian

[1] *Treatise on man; Passions of the soul.*

metaphysician. Descartes' rigid dualism was not acceptable to all philosophers and physiologists. Among the criticism delivered and alternatives presented was the truly rigorous conception of organic mechanism in which not only animal functions (movement, growth, reproduction) but also all intellectual functions, including the full spectrum of moral and mental behavior from the conscious to the unconscious, were indissolubly joined and then reduced for explanation to one fundamental substratum of being, matter. But matter was itself subject to varying definition. It was quite possible to argue, for example, that "matter" referred not to hard, impenetrable, insensitive, and passive particles as conceived by traditional atomism but presented in its definition a number of qualities, principally capacity-to-act and sensitivity, which brought it close indeed to the salient phenomena of "life," including both conscious and unconscious behavior. Such was the practice of the outspoken physiological materialists of eighteenth-century France. While their materialism occasionally verged on animism, and thus seemed to reintroduce from a different direction the perennial problem of the soul, commonly their objective was quite simply to exclude soul and its presumed properties from serious physiological discourse. With the soul held in abeyance (or denied, by some extremists, any reality at all), the student of animal and, most importantly, human function could get on with his ever-deepening analysis of the living machine.

A French philosopher-physician, Julien Offray de la Mettrie (*L'homme machine,* 1749),[2] had publicized the necessary relation between organic mechanism and the larger conception of philosophical materialism, and the argument remained a provocative and widespread one into the new century. It was destined, however, to be again vigorously questioned on both metaphysical and theological grounds. In the early decades of the nineteenth century, an era of acute spiritualism in religious belief and systematic idealism in metaphysics, the customary mechanistic-materialistic conception of the organism and particularly of man attracted more abuse than adherents. Many physiologists found the concept morally repugnant. Others felt it to be without evident applicability and value in the investigation of vital processes. For numerous and still quite insufficiently explored reasons the mechanistic ideal forcefully reentered physiology toward 1840. The experimental practice of physicist and chemist became the object of lively interest. Speculation

[2] *Man a machine.*

was renewed concerning the possibility that the explanatory concepts of these sciences, above all, those of mechanics and electrodynamics, might prove applicable to physiological phenomena. Historians have recently celebrated as decisive figures in these developments a small group of exceptionally able physiologists active in Berlin during the 1840s. This group, acting as prophets, proselytes, and propagandists of a presumably new and improved physiological science, catches the eye of hindsight but presents difficult problems for just assessment. In any case, the Berlin group (the so-called physiological "reductionists") may well have promoted a fresh sense of the possibilities of viewing the living organism as a machine.

That sense coincided with and was surely encouraged by the establishment of the doctrine of the conservation of energy. By 1847 diverse elements of this doctrine had been enunciated and systematized by Hermann von Helmholtz, James Joule, and a surprising variety of other workers, and concrete numerical values assigned to the conversion processes concerned. These processes had become increasingly evident during the nineteenth century. The steam engine daily demonstrated the conversion of heat into mechanical work. Current electricity, produced by chemical processes in the newly invented Voltaic pile, gave rise to heat and light. That same electricity, when suitably administered, brought about chemical dissociation. The interconvertibility of electricity and magnetism was demonstrated as was that of mechanical work and magnetism. These phenomena strongly suggested that most and presumably all physical processes were convertible one into another. Evidence pointed to a common basis for such conversions and many surmised that that basis (energy) might be expressible in precise numerical terms. From mathematical studies in dynamics, careful engineers' analyses of the efficiency of heat engines and the decisive experiments of Joule, a figure for the mechanical value of heat, subject to continual refinement, was obtained. Interacting physical processes could thus be precisely circumscribed by stating the energy relations of the changes produced. Equivalent amounts of energy would always produce equivalent amounts of electricity or heat or chemical change.

Of the authors of the conservation doctrine Julius Robert Mayer (1814–1878), trained in medicine and a practicing physician, most explicitly considered its bearing for our understanding of vital processes. Mayer's views were apparently little known but are nevertheless singularly illustrative of the potential contribution of the energy doctrine to physiology. Physics had shown, Mayer recalled, the existence of a determinate and steady mechanical equivalent of heat. Chemical combustion was a source of heat and thus energy values for chemical changes entered the relation. Now organisms were also sources of heat and motion and the bases for these phenomena must be found, Mayer claimed, in the fundamental vital chemical process, that is, oxidation, the ultimate source of energy for the living organism. "In the living body," he wrote (about 1852)

"carbon and hydrogen are oxidized and heat and motive power thereby produced. Applied directly to physiology, the mechanical equivalent of heat proves that the oxidative process is the physical condition of the organism's capacity to perform mechanical work and provides as well the numerical relations between [energy] consumption and [physiological] performance."

Mayer did not confuse ambition with accomplishment. Certainly the mechanical equivalent of heat was henceforth to be "in the nature of things the foundation for the elevation of a scientific physiology." But this new physiology lay in the future. Its coming would be slow; it would have to seize its rightful place in the scientific curriculum; it had yet to supplement and then most probably displace other modes of physiological investigation. Mayer had articulated, whatever was his influence, the grand objective of later nineteenth-century respiratory physiology. Animal heat had been shown to be the product of slow combustion. Net heat production, which would be recorded so as to include the heat equivalent of all bodily actions of chemical, electrical, or other nature, now was represented as the ultimate measure of the energy transformed by the organism. Arduous experimental researches were begun in the 1850s to prove that organisms do indeed act in full accord with the dictates of energy conservation; proof was delivered only by century's end. The organism in its overall measurable relations with the external world—that world serving as source and sink for the organism's energy supply—was an energy-conversion device, a machine no less than those scrutinized by mechanics and thermodynamics. This claim was justified despite the prominent fact that the animal machine's most intimate structure was still obscurely known and the nature of its essential intermediary chemical processes even more poorly appreciated.

Combustion, Respiration, and Life

Life, the modern physiologist argues, depends on the regular and slow release of energy derived from oxidation of ingested foodstuffs.

This energy provides the temperature appropriate to chemical reactions occurring within the body, including syntheses, and it underlies bodily motion, the electrical behavior of nerves and the secretory activity of glands. The certainty of this dependency, customarily advanced as a causal connection, derives from the accomplishments of nineteenth-century respiratory physiology. Recognition of the fact that heat appears in constant conjunction with life was made in classical antiquity, and is probably not unknown to most noncivilized men. This vital heat might be considered a continuous product of the organism or it could be regarded as inborn or innate, requiring at most and in some wholly unspecified manner the support of "air" from outside the body. The latter alternative was stated in antiquity and was still current, in related but more varied and suggestive idiom, in the eighteenth century. The essential fact requiring explanation was the same then as in antiquity, or, indeed, today. "It is clear," wrote Edward Rigby, a minor late eighteenth-century physiologist,

"that a very large proportion of heat must always be escaping from the body, and of course, such a quantity must be lost as to require the constant production of a considerable quantity in the body, to maintain the due proportion between them."

Rigby emphasized organic heat production, not its innate quality, and he presented a bizarre enough explanation (fermentive dissociation and mechanical attrition of foodstuffs) of that production. Already, however, a new view of animal (and plant) heat production was being cast and upon its central proposition, if not details, has arisen the entire modern science of respiratory and general physiology.

Antoine Lavoisier (1743–1794) first sketched his theory of respiratory physiology in 1777. Drawing upon the newly created chemistry of gases and, notably, recent researches on the role of oxygen in combustion Lavoisier, a leading figure in the chemical revolution of the eighteenth century, demonstrated that organisms disassemble and reconstitute atmospheric air in precisely the same manner as a burning body. That is, each active agent takes oxygen from the air and gives up to it carbon dioxide; neither utilizes the great store of nitrogen in the atmosphere. This discovery, supported by other workers, suggested a "much closer relationship" between respiration and combustion than might at first glance have been expected.

Further exploring this relationship Lavoisier, in collaboration with Pierre Simon de Laplace, introduced a capital instrument to physiologi-

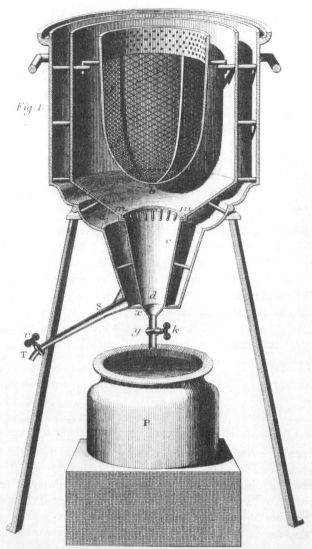

Figure 6.1 In the ice calorimeter, ice filled the space between the outer and inner solid walls and between the latter wall and the innermost, experimental basket. An experimental animal was placed in the basket and the entire apparatus then sealed and brought as far as possible into thermal equilibrium with the environment. Heat evolved by the animal would melt ice in the inner chamber and water thus produced was collected in the vessel "p." Such melt-water provided an index to the heat-producing capacity of the living organism, making the ice-calorimeter an indispensable instrument for early studies of the chemical and physical basis of respiration. (Antoine Lavoisier and Pierre Simon de Laplace, 1783 {1862}.)

125

cal investigation, the ice calorimeter. This device permitted estimation, by means of the amount of ice melted, of the quantity of heat evolved in a given period per unit of carbon dioxide produced. The figures obtained for a flame and for experimental animals (guinea pigs) showed reasonable agreement. Experimental and conceptual problems were numerous and the measurements subject to some doubt. Nevertheless, Lavoisier was satisfied that the data proved the essential point: with regard to heat production fire and life behaved in analogous and probably identical manner and the basis for both was the regular liberation of heat in the process of combustion (oxidation). "Respiration is therefore," he wrote (1780),

"a combustion, very slow to be sure but nevertheless strictly comparable to that of carbon. It takes place within the lungs . . . [and] the heat produced by this combustion is communicated to the blood which traverses the lungs and from there is distributed throughout the body. Thus the air which we respire serves two functions equally necessary for our preservation: it removes from the blood the base of fixed air [carbon] whose surabundance is quite deleterious and the heat which this combination [carbon plus oxygen] releases in the lungs restores the heat continually lost [by the body]."

Lavoisier's later researches further confirmed and extended the basic analogy.

While Lavoisier's physiological career ended abruptly with his execution in 1794 by the French revolutionaries, the example and conclusions set by his work lived on. On the one hand, Lavoisier's method, of which he was more symbol than unique representative, vastly reinforced the dreams and claims of those who sought fruitful application of chemistry and physics to physiology and perhaps all biology. Claude Bernard, for example, a master experimentalist of a later generation, could gladly assert in 1879 that the

"very essence of Lavoisier's doctrine lies in the affirmation that there are not two chemistries or two physics, the one applicable to living creatures and the other to inert bodies; rather, there are general laws applicable to all substance[s], however [they] might be disposed, and these laws admit of no exception."

On the other hand, Lavoisier's work truly defined the tasks facing respiratory physiology. His endeavor consequently presents a common point of reference for the manifold studies in this science during the nineteenth century. Lavoisier claimed that animal heat was the product of chemical combustion; that claim must be pursued and rigorously confirmed or denied. Lavoisier's suggestion that this combustion occurred in the lungs was to be explored and soon rejected; other sites within the body for heat production were proposed. The nature and source of the ultimate combustible substances (carbon, hydrogen) required attentive investigation. The precise manner by which oxidation in the body occurred became the object of hopeful inquiry and constant frustration. Finally, the full integration of the organism into the realm of physico-chemical parameters was attempted. From the study of respiration came recognition that the organism, too, was limited in its powers by the conservation of energy.

The Site of Respiration

Against Lavoisier's proposal that respiration (basically, that process which simultaneously produces carbon dioxide and liberates heat) was a form of combustion, serious objections were raised; combustion might destroy the lung tissue; the lungs were not appreciably warmer than other parts of the body. Already in 1791 a decisive new proposition was offered. The blood traversed the lungs not to be warmed and to convey that warmth to the body, but to accept oxygen. This oxygen then combined, in the blood and particularly within the capillaries, with available carbon and hydrogen. Heat was thereby gradually released and the carbon dioxide produced was brought by the venous system to the lungs and there discharged from the body. This was the principal model for understanding the respiratory-circulatory function until mid-century. The blood was dethroned from its high role only in the 1870s. Its place was then taken by the concept of tissue respiration, an idea narrowly bound to the functional attributes of the cell and one no less an exemplary product of physiology's ever-growing experimental adroitness.

It was William Harvey who sacrificed (1651) the ancient primary organs, heart and liver, in order to elevate the blood to physiological supremacy. His influence was still powerful in the late eighteenth century and was endorsed by the leading anatomist and physiologist of the period, John Hunter. To some, Hunter's notion of the "vitality of the blood" was the very cause of life and the blood the most truly living component of the body. On a less venturesome level one might at least claim that, since it received from the gut and transported throughout the body the prepared foodstuffs (chyle), the blood must be the site of essential physiological operations. Both views were wholly congruent with the idea that the blood was the site of respiration.

Experiment urged the same conclusion. Gustave Magnus' famous experiments, using a mercury evacuation apparatus capable of creating high vacuums and also employing techniques for introducing replacement gases into the blood, showed (1837) carbon dioxide and oxygen to be present in both the arterial and venous systems. The fact that the oxygen/carbon dioxide ratio was demonstrably higher in the arteries than in the veins proved that vital combustion could not take place in the lungs. It is important to recognize that Magnus neither proved nor claimed to have proved more than this. The inference was clear that respiratory chemical changes occurred throughout the body. Magnus admitted a preference for tissue respiration but noted that proof for this contention required exacting analysis of the blood gases and their changing proportions well beyond his or contemporary capacities. A respiratory role for the blood continued to be defended into the early 1870s, mainly in the work of Carl Ludwig and his students who argued (and offered apparent experimental confirmation) that tissue oxidations were indeed real but were subject to regulation by the chemical condition of the blood.

One group of observers began toward 1850 to perform experimental analyses on the respiratory activities of isolated preparations (usually muscle fibers). Georg Liebig showed that active muscle consumes oxygen and evolves carbon dioxide. If these two events are truly "respiration," then the muscle respires and can do so without connection to the blood. Helmholtz, in a superlative series of researches, had shown (1847) that an exchange of materials accompanied contraction of the muscle preparation. Moreover, and this was of fundamental importance to Helmholtz's study of energy conservation as well as to the more special investigation of animal energetics, Helmholtz detected thermal changes in the muscle. Muscular motion gave evidence of underlying chemical reactions and these were in turn evidently related to heat production. The relationship could be stated quantitatively.

The work of Georg Liebig and Helmholtz and of occasional others provided useful but unsystematic evidence for tissue respiration. Moritz Traube (1826–1894), a Breslau vintner and skilled physiological chemist, proposed in 1861 a concise "theory of the chemical processes of muscle action." Oxygen, he claimed, passed from the blood through the capillary wall and into the muscle fiber. There, in the fluidity of the cell, oxidation occurs and carbon dioxide and heat are produced.

Traube's claims were based on his earlier studies of fermentive reactions and not on the indispensable and as yet unavailable microanalyses of the blood gases. His theory was thus at once highly suggestive, in evident contradiction to the notion that oxidation takes place in the blood and desperately in need of friendly and reliable support. Moreover, Traube's work cast further suspicion on the then prevailing view of the chemistry of muscle action, proposed by Justus Liebig (see below).

Support for the idea of tissue respiration came in papers published in 1872 and 1875 by Eduard Pflüger (1829–1910), a versatile physiologist at Bonn, and from the continuing activities at Ludwig's Leipzig laboratory. Pflüger's 1872 paper is a landmark in the history of respiratory physiology. It was ascertained that even slight pressure differences across a cellular membrane could lead to a rapid and massive transfer of oxygen. Such small differences, Pflüger argued, are manifest throughout the body tissues and they, not the speed of blood flow, the presence of reducing substances in the blood or lung action regulate the rate of oxygen consumption in the body. Differences in the oxygen supply were themselves the direct consequence of metabolic activity within the tissues (cells). Oxygen was supplied from the blood in quantities appropriate to vital activity, this activity being essentially the breakdown of energy-rich substances available within the cells. The cell and its operations performed, therefore, the "essential animal work" and in so doing both controlled and depended upon a precisely regulated oxygen supply.

While apparently having drawn heavily upon Ludwig's researches Pflüger was assaulting the notion, supported for many years by Ludwig, that the blood and its gas content were the decisive regulative factors of the overall respiratory activity of the organism. Pflüger had the additional advantage of presenting his general review of the problem to a physiological public fresh from attending the downfall of Liebig's conception of chemical alterations in the body. The effect of these investigations—pursued by many and often conflicting physiologists and schools—was to ruin the physiological hegemony of the blood. The blood's role, insofar as regards the major features of respiration and energy production, was henceforth to be transportation. It brought oxygen (and foodstuffs) to the tissues and carried away the waste products of metabolic transformation. "For oxygen must go," Pflüger wrote, "everywhere in the body where there is life, and life is everywhere." Even the blood, of course, had its respiratory activity but this was of trivial magnitude in comparison to that of the great cellular masses composing the body. Experimental investigation, grounded in the assistance lent by chemistry and physics to the study of living membranes and body structures and fluids, had firmly established cell and tissues as the critical element in respiratory physiology.

A generation earlier Rudolf Virchow had also assaulted the hegemony of the blood, the primacy of body humors (see Chapter II). He had done so on pathological grounds and had sought to replace, by means of the cell doctrine, the conception of generalized disease with a new conception rigorously circumscribed by the essential functional element, the cell. That same element was now reliably designated the fundamental energy converter for the living organism. From these conclusions arose the plea for a truly "general physiology," that is, a study of vital processes common to plants and animals and one deemed general because it was grounded in the common denominator of all organisms, the living cell. Such was the appeal cast by Claude Bernard, the leading French physiologist of the period and a great publicist for his science. Bernard erected (1876 ff.) a grand synthesis of physiological thought on the basis of the essential and exclusive activity of the cells and of the fluids surrounding them. The cell was the principal functional component; it was surrounded by a nourishing and protecting internal environment. The general physiologist must explore both, gain mastery over the conditions affecting their behavior, and thus win deeper understanding of life and its distinctive phenomena.

The Nature and Source of Combustible Materials

Chemical analyses performed by Lavoisier and his French contemporaries had shown that the elementary composition of organic substances was primarily carbon, hydrogen, oxygen, and nitrogen. The study of the number and relative proportions of these elements in the composition of the constituents and products of living beings was the central theme of early organic chemistry. It was widely assumed that chemical transformations in the living and nonliving worlds were strictly comparable, the occurrence of such transformations being, of course, a fact given by observation and experiment. This evidence said little, however, of the actual transforming processes.

Swedish, French, and German chemists had perfected tools for chemical analysis during the first decades of the century. Following Justus Liebig's introduction (1831) of a simple and reliable apparatus for assaying elementary composition, organic analysis became a routine laboratory operation and a coherent roster of organic substances took shape. The division of foodstuffs into carbohydrate, fat, and protein, based on their proportions of carbon and hydrogen and on the presence of nitrogen, was firmly established by 1845. This knowledge, together with continued faith in the comparability of chemical transformations within and without the organism, opened interesting prospects for the study of physiological chemistry. In 1840 Liebig's *Agricultural chemistry* appeared, in 1842 the celebrated *Animal chemistry, or organic chemistry in its application to physiology and pathology.* With these works and the polemics, errors, and vigorous experimentation they immediately provoked, chemistry, after long having promised much to the student of organic function, now forced a massive entrance into physiological thought. From the position thus gained it has never been dislodged.

Animal chemistry, its author claimed, would "direct attention to the points of intersection of chemistry with physiology." Once defined, the physiologist would discover in organic, that is physiological, chemistry "an intellectual instrument by means of which he will be enabled to trace the causes of phenomena invisible to the bodily sight." Liebig (1803–1873), an able analytical chemist at Giessen, had already won a European reputation for the organization and effectiveness of his labortory-based teaching of chemistry. *Animal chemistry* was a synthetic statement of his views and also a public manifesto: let chemist speak directly to physiologist and be confident that both will benefit from the meeting.

Adopting the approach of contemporary French agricultural chemists, Liebig sought to establish an overall balance of the elemental constituents of the bodily ingesta, excreta, and respiratory gases. Liebig's premise was the conservation of matter, the basis of the new chemistry established by Lavoisier and his associates. Input must equal output and, fortunately, with regard to chemical transformations of the body, new methods of analysis promised a means for demonstrating that equivalence. Nevertheless, this overall elementary balance of the foodstuffs, oxygen, and the solid and fluid wastes was only Liebig's preliminary target. He hoped to be able to deduce "phenomena invisible to bodily sight," and the phenomena in question were the actual processes of chemical transformation that occurred within the body. The physiological chemist could control and assay only the elementary balance of an experimental organism; now he must use this data to divine necessary intermediary reactions.

Subsequent criticism demonstrated the futility of this approach. One's

PROGRAMME OF THE DISCOURSE

AN ANIMAL		A VEGETABLE	
is		is	
An Apparatus of Combustion;		An Apparatus of Reduction;	
Possesses the faculty of Locomotion;		Is fixed;	
Burns	Carbon, Hydrogen, Ammonium;	Reduces	Carbon, Hydrogen, Ammonium;
Exhales	Carbonic Acid, Water, Oxyde of Ammonium, Azote;	Fixes	Carbonic Acid, Water, Oxide of Ammonium, Azote;
Consumes	Oxygen, Neutral Azotised matters, Fatty matters, Amylaceous matters, sugars, gums;	Produces	Oxygen, Neutral Azotised matters, Fatty matters, Amylaceous matters, sugars, gums;
Produces	Heat, Electricity;	Absorbs Heat, Abstracts Electricity;	
Restores its elements to the air or to the earth;		Derives its elements from the air or from the earth;	
Transforms organized matters into mineral matters.		Transforms mineral matters into organic matters.	

Figure 6.2 The investigation of the nutrient balance of living creatures and of other factors required in the organism (for example, heat or electricity) led to study of the metabolic process and also to the recognition of essential cycles of physical and chemical exchange between the inorganic and organic realms and between plants and animals. One too easily devised, however, a perfect symmetry between the physiological capacities of the plant, supposedly the exclusive synthetic agency, and those of the animal, which received its "organic matters" from the plant, destroyed them and returned the resulting simple chemical substances to the environment. By the 1850s the vast complexity of metabolic processes was becoming apparent and assertive conclusions, such as those given in this Program, disappeared from physiology. (Jean Baptiste Dumas and Jean Baptiste Boussingault, 1844.)

data pertained exclusively to initial and terminal conditions and, given the limited power of contemporary chemical theory and the dearth of relevant experience, it offered no real grounds for inferring the nature of intermediary reactions. Liebig recognized these limitations and in later statements moderated his claims. The idea, however, had taken firm hold and was not to be discredited until the 1860s.

The doctrine itself is of considerable interest. Liebig intended to relate specific vital activities to specific foodstuffs. Carbohydrate and fat upon oxidation in the body produced the heat of life. The organic molecules were ultimately wholly reduced to carbon dioxide and water and these products expelled from the body. Carbohydrate and fat were, in Liebig's terminology, the "elements of respiration." But no less interesting to the question of organic structure and action were the "plastic elements of nutrition," that is, the proteins, which Liebig believed constituted the blood and flesh itself, particularly muscle tissue. The muscles were responsible for animal motion, the most distinctive quality of the beast, and the molecular basis for this motion was clearly to be the nitrogenous materials of muscle (proteins). Protein was broken down and yielded carbon dioxide, water, and the important nitrogenous substances, principally urea, found in the urine. It also yielded "force," the agency ultimately responsible for animal motion. "All vital activity," said Liebig, "arises from the mutual action of the oxygen of the atmosphere and the elements of the food." The most obvious form of vital activity is motion and motion is due to muscular action. Liebig claimed that ingested protein was directly assimilated into muscle tissue, the source of all protein being the synthetic activity of plants. Protein degradation alone supported muscular activity. Carbohydrates and fats produce only heat and "protect the organism from the action of atmospheric oxygen," meaning that only nitrogenous substances are part of the fundamental structure or tissues of the organism.

If, therefore, protein degradation accompanied and perhaps caused true animality, that is, motion, the degradation products—above all, urea—offered an exact index to animal activity. Here was the critical point both for Liebig and his critics: nitrogen excretion, when taken as the measure of muscular work, must vary directly with the work output of the animal. Facts readily showed that such was not the case. Traube (1861) pointed out that the great working animals (horse, ox, camel) are herbivores. The busy bee, moreover, subsists largely on sugar. Muscular activity seemed, indeed, more closely related to oxygen consumption and carbon dioxide production than to a (doubtful) rise in nitrogen excretion.

Liebig's scheme was severely tested and effectively destroyed in 1865–1866. By this time physiologists freely employed the idea of energy utilization in analyzing organic activities. Energy offered a common basis for evaluating the work output of an active animal and the total heat, thus energy, available in any or all foodstuffs that an organism might consume while working. Two German physiologists, Adolf Fick and Johannes Wislicenus, had measured their nitrogen excretion while climbing an Alpine peak. Computing the energy available in the protein presumably consumed by their exertions and comparing this with a reasonable estimate of work actually performed they found the former quite inadequate for its designated purpose. Their data, nevertheless, were approximate and their conclusions subject to criticism. Such was not the case with the devastating assault on the protein-degradation theory of animality delivered by Edward Frankland (1825–1899). Frankland, a London chemist trained in Germany, employed a bomb calorimeter, carefully correcting for possible losses of heat from the apparatus. His experimental data allowed him to compare the mountain climbers' energy consumption (conservatively estimated) and a precise value for the energy released by total combustion of a quantity of muscle tissue equivalent to the nitrogen excreted. His conclusions follow:

	Energy available	Work performed
Fick	68700 KgM	160000 Kgm
Wislicenus	68400 KgM	184000 KgM

This startling discrepancy indicated that the energy available in muscle tissue alone could hardly account for the work actually performed. This is all the more true, Frankland stressed, when it is recognized that muscle is considerably less than 50 percent efficient; that is, its energy is by no means fully available for muscular work. Clearly, Liebig's belief that muscle tissue (protein) was the exclusive source of animal activity was false. The muscle fiber was indeed the part that moved, but the energy for that movement must derive largely from other sources, that is, from foodstuffs other than protein. "A command," Frankland concluded,

"is sent from the brain to the muscle, the nervous agent determines oxidation. The potential energy becomes actual energy, one portion assuming the form of motion, the other appearing as heat. *Here is the source of animal heat, here the origin of muscular power!* Like the piston and cylinder of a steam-engine, the muscle is only a machine for the transformation of heat into motion; both are subject to wear and tear, and require renewal; but neither contributes in any important degree, by its own oxidation, to the actual production of the mechanical power which it exerts."

His metaphor is common and Frankland's view of muscle physiology too simple, but the ramifications of his sharply pointed conclusion are profound.

Frankland's work culminated the destruction of Liebig's protein theory. In so doing it reemphasized how perilous it was to seek to deduce intermediary metabolic processes by an examination of external chemical phenomena. Frankland's superb experiments urgently suggested that the muscle is a machine for energy conversion. If they appear to prove this special point they certainly could not demonstrate that the organism as a whole might be regarded as a heat engine. Nevertheless, Frankland's technique—complete combustion analysis of tissue and foodstuffs—and general orientation clearly defined the approach later to be used for attaining that demonstration.

The Perpetual Dilemma: Biochemical Reactions

Frankland's conclusions also suggest the frustration facing those who hoped to erect a dependable science of plant and animal chemistry. It was one thing to analyze the chemical products of living bodies; it was quite another to grasp the nature of the reactions that produced these substances. Lavoisier's example, perpetuated in virtually unmodified form by Liebig, assigned to oxygen a preeminent role in physiological chemistry. Oxygen was the active agent of change. It attacked combustible substances and its supply appeared to regulate the role of respiratory chemical reactions within the body. "Truly, the history of oxygen," Traube exclaimed, "is the history of life!" Only in the 1860s and 1870s, mainly through the assignment by Ludwig, Pflüger, and others of basic respiratory processes to the cell and its constituents and through Voit and Pettenkofer's investigation of the regulatory role of oxidizable substances (see below), did this simplistic notion of organic combustion diminish.

But the larger question remained: What indeed was the nature of combustion in living creatures? Complex and usually stable molecules were broken down in the body. Under atmospheric conditions these same molecules remained unaffected, but within the body and there-

Weight and Cost of various articles of Food required to be oxidized in the body in order to raise 140 lbs. to the height of 10,000 feet.
External Work = one-fifth of Actual Energy.

Name of food.	Weight in lbs. required.	Price per lb.		Cost.	
		s.	d.	s.	d.
Cheshire cheese	1·156	0	10	0	11½
Potatoes	5·068	0	1	0	5¼
Apples	7·815	0	1½	0	11¾
Oatmeal	1·281	0	2¾	0	3½
Flour	1·311	0	2¾	0	3¾
Pea-meal	1·335	0	3¼	0	4½
Ground rice	1·341	0	4	0	5½
Arrowroot	1·287	1	0	1	3½
Bread	2·345	0	2	0	4¾
Lean beef	3·532	1	0	3	6½
Lean veal	4·300	1	0	4	3½
Lean ham (boiled)	3·001	1	6	4	6
Mackerel	3·124	0	8	2	1
Whiting	6·369	1	4	9	4
White of egg	8·745	0	6	4	4½
Hard-boiled egg	2·209	0	6½	1	2½
Isinglass	1·377	16	0	22	0½
Milk	8·021	5d. per qrt.		1	3½
Carrots	9·685	0	1½	1	2½
Cabbage	12·020	0	1	1	0¼
Cocoa-nibs	0·735	1	6	1	1¼
Butter	0·693	1	6	1	0½
Beef fat	0·555	0	10	0	5½
Cod-liver oil	0·553	3	6	1	11¼
Lump sugar	1·505	0	6	1	3
Commercial grape-sugar	1·537	0	3½	0	5½
Bass's pale ale (bottled)	9 bottles.	0	10	7	6
Guiness's stout "	6¾ "	0	10	5	7½

Figure 6.3 Energy provides an index to capacity for doing work. The working machine may well be the human frame, the energy source being the foodstuffs fed the worker. This interesting table, prepared by a chemist with close Manchester and London industrial connections, provides a novel measure of foodstuff value, cost in pounds, shillings and pence per energy unit produced. While there is no discussion here of the needs of a balanced diet, there is a clear suggestion to the cost-conscious industrialist of how wages might be neatly weighed against the essential energy requirements of his laborers. (Edward Frankland, 1866.)

fore at relatively low temperature and giving no sign of flame or effervescence these molecules were rapidly and easily destroyed. Systematic and often successful investigation of biochemical reaction processes belongs to the twentieth century. Useful clues, however, were becoming known during the previous century. The idea of catalysis was introduced to physiology by Jöns Jacob Berzelius in the 1830s. He argued that in organisms doubtlessly thousands of reactions were furthered by chemical agents (catalysts) whose presence was essential to a reaction but which were not consumed in the reaction. In the same decade and increasingly thereafter peculiar biological agents (ferments) were identified. These complex substances of unknown constitution acted like catalysts and at body temperature mediated many intricate chemical transformations, including, it was later found, those which liberated energy as heat or in a bound form usable by the organism.

Many ferments (for example, pepsin, discovered in 1836, or diastase, discovered in 1833) were isolated and understood to be organic products. The expression ferment was of long standing, however, and it had been and continued to be applied not only to various well or poorly specified organic reactants but to the essential agent inducing the familiar process of alcoholic fermentation. In fermentation a simple sugar (glucose) produces alcohol and carbon dioxide and energy is freed. Berzelius, Liebig, Traube, and other chemists attributed alcoholic fermentation wholly to molecular catalytic action, that is, to strictly inorganic agencies. Louis Pasteur (1822–1895) disagreed and championed the idea that fermentation was essentially a physiological or "vital action," that is, a chemical process necessarily "associated with the life and organization of yeast cells." The opposition between the two views—the chemists' catalyst and the physiologists' cell—later proved illusory. Edvard Buchner in 1897 separated a ferment (zymase) from the yeast cell that produced alcohol and carbon dioxide from glucose, and did so in total isolation from any cell. Ferments, or enzymes (a term introduced in 1877), now appeared to be true physiological, that is, cellular products. Their activity, however, seemed independent of their site of formation; they were genuine chemical catalysts.

Certain other vital work should be added to this brief record of nineteenth-century contributions to the study of respiratory processes. Pasteur's discovery (1861) of anaerobic fermentation disclosed, for example, that heat production within the cell did not require the presence of oxygen, and thereby further diminished the exclusive physiological rights of that element. Felix Hoppe-Seyler, a German physiological chemist, and George Stokes, a distinguished British physicist who followed Hoppe-Seyler's lead, applied (1862, 1864) the newly invented spectroscope to the blood pigments and confirmed Liebig's suspicions regarding them. Iron compounds, Stokes noted, impart various colors to the blood and also served as "a greedy absorber and carrier of oxygen." Much later it was found that the binding of oxygen to these substances is easily reversible and thus one could propose these substances in the blood as the vehicle conveying oxygen to the tissues.

The cumulative effect of these discoveries was important but cannot be said either to have resolved the critical issues of respiratory physiology or to have created a coherent science of physiological chemistry. The diversity of chemical reactions within the organism, the unparalleled technical difficulties impeding their study and the evident imperfections in knowledge regarding catalysis and the nature of biochemical reactions in general powerfully discouraged great expectations. Frederick Gowland Hopkins, creator of the noted Cambridge University school of biochemistry, reviewed (1913) the history of his science and found it wanting. A new faith, he recognized, was needed and he stated it in unequivocal terms:

"[I]n the study of the intermediary processes of metabolism we have to deal, not with complex substances which elude ordinary chemical methods, but with simple substances undergoing comprehensible reactions. . . .[I]t is not alone with the separation and identification of products from the animals that our present studies deal; but with their reactions in the body; with the dynamic side of biochemical phenomena."

Hopkins, his colleagues, and contemporaries acted on this faith and soon moved physiology from the observation of general cellular operations to the experimental analysis of intracellular chemical and physical mechanisms.

Calorimetry and the Organism as a Machine

Liebig's study of the elementary chemical balance in animals had broader aims than mere physiological theory. He and his contemporaries were deeply concerned with the nutritional value of foodstuffs, that is, the contribution of each article in the diet to the healthful vitality of the organism and the establishment of a basic normal maintenance ration. This interest probably first arose from the study of soils and the needs of what was regarded as a new, scientific agriculture and husbandry. But the applicability of these nutritional studies to man and human society seemed obvious. During the 1840s and 1850s efforts were made to extend the knowledge of nutrients to public health measures and general social amelioration. Chemical analysis of foodstuffs proved, however, to have promised more than it could deliver. Foodstuff analysis soon blended with food fadism—perhaps most notably in Liebig's beef extract (held to be uncommonly rich in restorative protein) and Sylvester Graham's famous whole wheat crackers—and not until century's end, with the study of deficiency diseases and the discovery of vitamins, did this aspect of nutritional studies revive.

It seems no less true that the investigation of specific nutritive values was quite overwhelmed by the fresh possibilities inherent in the assessment of the energy values of foodstuffs. From 1860 onwards vital energetics becomes the fundamental theme in respiratory and metabolic studies. Its classic instrument was the respiratory calorimeter. The heat produced by the living body could be measured either directly or indirectly. The former approach was exemplified by Lavoisier and Laplace: the quantity of ice melted in their calorimeter was regarded as a direct index of heat evolved by the experimental animal. Indirect measurement of animal heat production was more complicated. Independent experiments (for example, those of Frankland) demonstrated that carbohydrate, fat, and protein each produced, on total combustion, a definite quantity of heat and distinctive gaseous products in regular quantities. From the latter could then be calculated the quantity of foodstuff(s) broken down in the body. This quantity could in turn be converted into heat equivalents. In organisms, however, the final products of respiration are not wholly gaseous. Some nitrogenous substances (urea, uric acid, creatinine) are expelled in the urine. These products are not totally oxidized and therefore remove from the body a certain quantity of energy introduced in the foodstuffs. The indirect method requires, therefore, scrupulous attention to all respiratory products, be they lost to the body by the lungs, the skin, the bladder, or the bowels.

The basic apparatus for indirect measurement of heat production was introduced in 1849. It was a closed system of tubes and chambers. Carbon dioxide produced by the enclosed experimental organism would be absorbed and weighed; measured quantities of oxygen could be introduced as required. Constant improvements were made on this device, most notably that of providing a constant flow of precisely controlled air into the apparatus. Gases leaving the instrument could be sampled at regular intervals and thus experiments of relatively long duration performed. Using this ventilated or open circuit and collecting solid and fluid excreta Carl Voit and Max von Pettenkofer in their Munich laboratory carried out important investigations on respiratory changes. They demonstrated (1862) that oxygen utilization varied with the nature of the foodstuffs consumed. Thus, not oxygen supply but the availability of breakdown products from the foodstuffs controlled the respiratory rate. This was a telling blow to those who had overemphasized oxygen's direct role in the body and it compelled much closer attention be paid the physiological condition of the body (health, hunger, rest).

While Voit and Pettenkofer thus perfected the respiration apparatus they paid little attention to calorimetric analysis. The techniques of direct and indirect measurement of animal heat production were ultimately combined by Voit's student, Max Rubner (1854–1932). Rubner himself pointed out that it was scarcely necessary to propose to the scientific world that the conservation of energy applied to the biological realm. That applicability had, however, either been assumed or only loosely tested; Rubner's classic experiments (1889–1894) were designed to put this widespread belief beyond all question.

Rubner placed a ventilating respiratory apparatus within a calorimeter, thus forming a respiration calorimeter. This combination permitted the simultaneous recording of the respiratory gas exchanges and direct measurement of heat production; the nitrogeneous wastes in the urine had to be collected and their heat value established. Rubner exerted exceptional effort to correct all errors in his apparatus, to command, as he remarked, "all biological factors" pertaining to the investigation. This unrelenting thoroughness was doubtlessly the condition of his achievement, for the design and objectives of his experiments could not in the 1890s be considered novel.

The purpose of the experiments was to consider "whether the substances burned in the body possess the same quantity of heat as was given off by the surface of the animal body." Rubner measured the heat production of large and small dogs placed in the respiratory apparatus. He varied the diet (fasting, pure fat, fat and meat, pure meat) and recorded the ensuing shifts in the respiratory relations. Throughout these numerous experiments Rubner discovered only the closest agreement between the values for the quantity of heat produced by a given animal as measured by the indirect and direct methods (see Table 1). Indeed, the slight variation of this figure fell within the margin of error

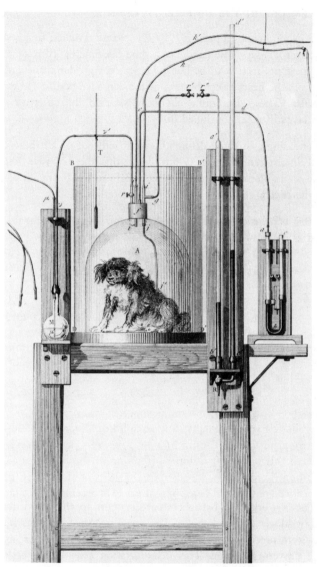

Figure 6.4 The first enclosed respiration apparatus is shown here. Oxygen is supplied to the dog by the tube connected to the left; carbon dioxide produced by the animal is removed from the experimental chamber by the tubes at the upper right. The chamber is fitted with instruments to measure temperature and pressure variations. With this device and subsequent improvements made upon it, it proved possible to control and assay, over extended periods of time, the consumption and evolution of virtually all gases of physiological importance. (Henri Victor Regnault and Jules Reiset, 1849.)

of the experimental apparatus. "In the overall average of all experiments over 45 days," he announced, "the calorimetric [direct] method shows heat production to be but 0.47 per cent greater than was calculated from the heats of combustion of the degraded bodily and nutritive materials." These data quite obviously satisfied the objective of the experiments. Rubner concluded that

"Not a single, isolated datum chosen at will out of all these experimental results can leave us in any doubt that the exclusive source of heat in warm-blooded animals is to be sought in the liberation of forces from the energy supply of the nutritive materials."

That this conclusion would not equally apply to other animals he regarded as an "absolutely incomprehensible hypothesis."

Table 1

Diet	Duration (Days)	Calculated Heat [Indirect]	Measured Heat [Direct]	Difference (%)	Average Difference (%)
Hunger	5	1296.3 Cal.	1305.2	+0.69	−1.42
	2	1091.2	1056.6	−3.15	
Fat	5	1510.1	1495.3	−0.97	−0.97
Meat and Fat	8	2492.4	2488.0	−0.17	−0.42
	12	3985.4	3958.4	−0.68	
Meat	6	2249.8	2276.9	+1.20	+0.43
	7	4780.8	4769.3	−0.24	

Rubner's work thus brought secure support to the conviction first stated by Mayer and Helmholtz. The organism was a heat machine. Its forces were precisely comparable and most probably the same as those of the universe at large. This demonstration contributed, of course, nothing to our knowledge of the intermediary reactions within the body, those reactions whose net effect was alone measured by respiration calorimetry. But many physiologists believed the demonstration set limits, necessary yet promising limits, to biological investigation. "The most striking characteristic of living organisms," wrote William Bayliss in 1913, "is the perpetual state of change [or work] which they show." He continued: "This capacity of doing work is due to the possession of something which is called energy." Energy, however, is the mark of physical and chemical processes and it is these alone with which the experimental physiologist can cope. Bayliss' book (*Principles of General Physiology*) was designed to introduce a systematic rethinking of the structure of physiology, and the keystone to that structure—energy—was derived from the above pronouncements. "It must be kept in mind," Bayliss demanded,

"that all the methods available for the study of vital processes are physical or chemical, so that, even if there were a form of energy peculiar to living things, we could take no account of it, except when converted into known forms of chemical or physical energy of equivalent amount."

Bayliss was referring to the larger question of confining vitalistic explanation but he also exposes the conceptual power within biology of the doctrine of the conservation of energy. Above all, that doctrine provided concrete points of reference for comprehending the overall respiratory and metabolic changes within organisms. Study of the exchange of gases and the heat value of foodstuffs and excreta united about the term energy. And the conversion of the potential energy of the nutrients into motion and into energy usable for other metabolic processes provided an outline (if not the details) of how animal heat was produced. This was far more than simply denying that such heat was inherent in the living creature and, indeed, it subsumed Lavoisier's conclusion that life was maintained by a slow combustion. The animal machine was still mysterious in its working parts but its general place in the universe—discounting the unresolved problems of psychic activity—seemed henceforth clear. The organism was no being apart, but an integral, interacting element in the physical universe.

Function the Object of Physiology

The "actions of living matter," wrote Thomas Henry Huxley in 1875, "are termed its *functions*." Thus, respiration and its associated chemical changes were functions, and so were the multitude of other organic actions—nervous transmission, digestive chemistry, cardiac movement and blood flow, glandular secretion—carefully analysed during the nineteenth century. Organisms, Huxley argued, are more than natural bodies manifesting definite structure and reproductive behavior. They are "living machines in action; and, under this aspect, the phenomena which they present have no parallel in the mineral world." Physiology becomes thereby the science whose special responsibility it is to study functions, the separate vital mechanisms of the organism as well as their collective effect, life itself.

But physiology was long a science in search of a method. In truth, physiology had possessed since antiquity a spectrum of methods for investigating organic functions. These methods—principally, observation and comparison, morbid anatomy, vivisection and, a much later addition, systematic physico-chemical experimentation—each had their partisans who were inclined to advocate the exclusive rights of their preferred procedure. Popular opinion easily transforms these alternatives into a progressive sequence, equating progress in the physiological sciences with increasing utilization of experimental techniques. This conclusion is, of course, too simple and too loosely drawn. It does not distinguish, as numerous nineteenth-century physiologists were wont to do, varieties of useful experimental procedure and, more importantly, it fails to indicate the decisive interplay between the physiologist's adopted methods and his conception of life and the organism. The latter commonly determined or, at the very least, offered essential premises for the determination of the former. The understanding of life and organism and the suitability of one's means of investigation stand inseparable.

In this context brief note must be taken of physiological instruments. The demonstration that living organisms were heat machines subject to the conservation of energy depended on the Lavoisier-Laplace ice calorimeter, Liebig's apparatus for the analysis of organic molecules, closed and open respiratory-gas analysis chambers, the spectroscope, and countless other physiological devices. But the design and construction of these instruments also depended on the lively conviction of their creators that the data that these devices could precisely record would have physiological meaning. Implicit in the instruments was an interpretation of life. While the story of the study of respiration during the nineteenth century lends itself well to this conclusion—bearing in mind that that story has been related here in a highly selective fashion which obscures the uncertainties, false directions, and outright errors characteristic of work in every science—developments elsewhere in physiology lend supporting evidence. By no means was an identical view of life necessarily implied or postulated by investigators of respiration, of embryonic development, or of the neural or secretory foundations for the harmonious maintenance of the functional equilibrium of the normal organism. Nevertheless, methods and instruments suitable to the problems of each of the expanding biological specialities were indeed devised, and most took guidance and the charmed word, experiment, from physiology.

This is a familiar conclusion but its importance deserves repetition. Bayliss' remarks on the apprehension of various forms of energy may be expanded. Our instruments, the outward signs of our science, can record information only from the realm for which they were designed. Their design and, perhaps more importantly, the data they return are essential physical and intellectual components of the science. They are not, of course, the entire science, for that will include an extended intellectual superstructure of varying scope and coherence, satisfying fitness with nature, and wealth of suggestion for future inquiry. Principally within this larger domain will be decided the terms and limits to be observed in serious consideration of the meaning of life and organism; of these terms physiological instruments are consequently an eloquent expression.

To some physiologists the conclusions of respiratory research suggested, perhaps demanded, that the organism was, in a wholly literal sense, a physical or chemical machine. In the nineteenth century such language led directly to mechanics, presumably the most fundamental of all sciences. Life and organism might, therefore, be reduced to peculiar patterns of matter in motion, or to perhaps an even more fundamental controlling force or forces of nature. Other physiologists saw no need for this conclusion. The fact that organisms could be analyzed by physico-chemical means was a valuable addition to physiological method but it need not entail or even suggest any conclusions regarding the essence itself of life. To such men this was a metaphysical question and consequently one not to be resolved by the methods and conclusions of physiology.

These philosophical positions are exemplified by other areas of nineteenth-century physiological concern. The former or reductionist view reawakened the traditional materialistic interpretation of life. The latter viewpoint concentrated on vital phenomena and their relations and shunned rigid pronouncements on the ultimate nature of things. This was possibly the most common conception of physiological science entertained during the nineteenth century and had probably become the predominant one by century's end.

Yet another option was available, that of vitalism. No expression in the language of biology is so ambiguous and open to misuse or abuse.

That vitalists were numerous in the nineteenth century and did wonderful things with their "vital forces" is an indisputable fact. It is not at all clear, however, what this activity meant, what were the qualities and capacities of one or another vital force and what was the overall development of vitalistic thought during the period. In the early decades of the century, nevertheless, the issue of vitalism appeared to play a significant role in the emerging science of organic chemistry. Organic substances, it seemed, were formed only in organisms and were thus a product of that same vitality which maintained the living being. At the center of this discussion were the chemists Berzelius, Friedrich Wöhler, and Justus Liebig.

Whatever be the manner of interpretation, these distinct forms of physiological doctrine obviously agreed that organic function was the central problem of the science. Yet, the interpretations—mechanism, vitalism or otherwise—alone can provide some sense of the solutions brought forward to resolve the problem of function and, by extension, of the nature of life. Our ignorance of these matters remains, regrettably, large. No available study offers a comprehensive and reliable account of the evolving methodological interests and explanatory ideals of nineteenth-century biology. The analysis of special topics is equally deficient. The following discussion of physiological method and understanding offers, therefore, only suggestions and illustrations, and pretends to provide no full or indisputable account of these interests. The exposition will be confined to three topics, each subject to historical investigation in recent years: chemical vitalism, physiological reductionism, and the perhaps more temperate or positivistic stance which proclaimed ignorance and scientific disinterest before questions of essence or ultimate being.

Vitalism and Organic Chemistry

In 1828 Friedrich Wöhler (1800–1882), a young Berlin chemist, announced the laboratory synthesis of urea. Urea was an organic substance and its production, like that of other such substances, was commonly held to be the exclusive prerogative of the organism. The organism obviously must possess a special capacity or, indeed, an irresolvable vital force by which to effect these otherwise incomprehensible syntheses. Wöhler's synthesis has long been pronounced the death blow to vitalism and the real opening of the organism to chemical analysis. Recent scholarship, however, has denied the premises of this assumption and indicated quite another and essentially chemical meaning for Wöhler's accomplishment. According to this interpretation Wöhler contributed important evidence bearing on the naissant study of chemical isomerism. Isomers are substances with the same atomic composition but different chemical and physical properties. Urea and ammonium cyanate shared the empirical formula, CH_4N_2O, and were consequently isomeric. Wöhler's synthesis was held, moreover, to throw new light on particular problems of organic chemical combinations.

Neither of these contributions had obvious physiological bearing and neither, of course, spoke to the existence or nature of a vital force. Apparently it has been the historian and not the chemists of the 1820s and 1830s who have regarded organic synthesis as the critical issue for a vitalism based on the chemical capacities of the organism. Most chemists of the period believed that artificial syntheses, that is, production of organic substances outside the living body, were possible. Their likelihood would depend only on mastering the great technical difficulties encountered. Berzelius (1779–1848), a leading teacher and expositor of chemistry to the new generation, came to believe in the essential equivalence of inorganic and organic chemical substances and processes. Organic substances were simply more complex in structure and production, and were not different in kind. "The starting point" of life, Berzelius wrote in 1806,

"should be sought in the basic forces of the elements and it [life] is a necessary consequence of the relationships whereby the fundamental materials are joined . . . Consequently, there is no special force exclusively the property of living matter which may be called a vital force or force-for-life; rather, this force arises from the conflict of numerous other [forces] and organic nature possesses no laws other than those of inorganic nature."

To this opinion Berzelius apparently remained true throughout his career. But elsewhere Berzelius does refer to and emphasize the role of a vital force in the affairs of the organism. In 1827 he spoke of a vital "something-other," differing from the "inorganic elements" and not sharing their primordial qualities. The reference was not, however, to a controlling factor in chemical combination. What Berzelius seems to have in mind was a regulative force which brought forth "organization," that singular combination of tissues and organs that is called the organism.

Berzelius thus restated (see Chapter III) the need for a developmental force, whose abstract definition he wisely avoided but which surely controlled the formation of a new living being. Once the organism had been fully formed or, in Berzelius' expression, once "the relationships whereby [its] fundamental materials are joined" are attained, an adult organism exists; its principal task is that of maintaining a prosperous concern. To Berzelius and to hosts of other "vitalists" the production of organic form—individual development—was the primary sporting ground for the vital force; the problem of the regulation of on-going physiological processes in the mature organism offered far less invitation to play. Maintenance processes seemed explicable, at least to the more optimistic physiological chemists, on chemical grounds. Berzelius' was thus a residual vitalism. It defined no separate vital force but reserved that term for events (mainly embryonic development) still and perhaps permanently outside the accepted sphere of explanation for other bodily processes. This conclusion may require some qualification: Berzelius was a theist and regarded all creation as God's peculiar handiwork. His use of a vital force may have had divine, if not physiological sanction. However this be, it can be seen that Wöhler's synthesis, far from being the turning point in physiological chemistry and the exorcism of vitalism, was indeed a triumph for the chemists but was largely irrelevant with regard to the chemistry of life.

Functions other than reproduction and development might be assigned to a chemical vital force. Justus Liebig, the century's early master of physiological chemistry, explained the assimilation of foodstuffs by the action of such a force. Chemical and vital forces appeared to act antagonistically: "The chemical force which kept the elements [contained in the food] together acted as a resistance, which was overcome by the active vital force." The shifting dynamism of life, manifested above all in the varying intensities of tissue formation and degradation, becomes the consequence of a ceaseless interplay between the chemical and vital forces.

It would appear that these and similar declarations suffice to identify Liebig as a characteristic vitalist. Such a conclusion requires, however, severe modification and perhaps might fairly be altogether denied. Liebig's specification of the vital force is full and carefully considered. His was, like that of Berzelius, a residual vitalism. Many organic phenomena could be resolved by chemical analysis and modes of explanation; many others had yet to receive such concrete foundations. Among the latter were, for example, assimilation, growth, and bodily motion.

Liebig spoke of chemical forces as well as the vital force. Nevertheless, both exhibited a common and fundamental property: they were dependent on the "order in which the elementary particles [of a body] are united together." To destroy organic form, the structural and functional organization of the body, is to destroy life itself. "There is," Liebig consequently asserted (1842),

"nothing to prevent us from considering the vital force as a peculiar property, which is possessed by certain material bodies, and becomes sensible when their elementary particles are combined in a certain arrangement or form. This supposition takes from the vital phenomena nothing of their wonderful peculiarity; it may therefore be considered as a resting point, from which an investigation into these phenomena, and the laws which regulate them, may be commenced.

The vital force is, in short, a "peculiar force, for it exhibits manifestations which are found in no other known force."

Of primary importance in these remarks are the words "manifestations" and "known." Liebig invoked the vital force more as expression than entity and did so to cover obvious activities of the organism which otherwise might lack any explanation. The fact that every effect must have its cause was his declared premise, and to Liebig such causal explanation must be stated in terms of "force." Just as the motion of a falling weight is a result of gravitation and that of a flexed spring is a result of elasticity so must vital "motion" (assimiliation and resistance to external agents) be explained by a "vital force."

Liebig's interest in the vital force, an interest much remarked by his contemporaries, was one that aimed principally at vital phenomena and less at the delineation of a singular force that virtually by definition would be incommensurate with the phenomena it was called upon to explain. Liebig was not a metaphysician but a chemist with a profound interest in vital phenomena. While his vital force was indeed a "peculiar force" its effects were "regulated by certain laws" and these "laws must be in harmony with the universal laws of resistance and motion." So long as cause-and-effect explanation in this simplistic form prevailed, so long as vital phenomena could not—whether as a result of our ignorance or of the conviction that life is truly an irreducible phenomenon —be subsumed by the explanatory categories of other and presumably more fundamental·sciences, just so long might the varieties of vitalism prosper. Vitalism is always a possible mode of explanation; only its utility and contribution to the scientist's intellectual satisfaction are in question. The example of Berzelius and Liebig illustrates an aspiration for comprehensive explanation of biological processes. These chemists and those who shared their views were sustained by a faith that chemical principles were widely applicable to the organism and its behavior. But of course not all such behavior—generation, assimilation, and so on—could receive chemical explanation. Nevertheless, total explanation was demanded, loose ends were disturbing and undesirable, and perhaps hidden truths were still to be uncovered. To such recalcitrant phenomena vitalistic explanations continued to be easily applied.

A Physicalist Option

In claiming force as the operative form in biological explanation Liebig had employed the language of the German nature-philosophers (see Chapter III). He continued as well their claim on absolute explanation, that is, the demand for a fully articulated system of thought based on a minimal number of carefully selected premises. The same behavior characterizes the thought of the physiological reductionists. They retained force among their premises and with it brought about the motion of brute matter. In force and matter-and-motion they found explanatory satisfaction. This was a new, radical, materialistic generation which, in philosophical terms, sought to effect a revolution in ontologies, reassigning priorities so as to give the principles of mechanics preponderant influence throughout the sciences.

The principal reductionists were German. During the 1840s criticism of political absolutism and religious conservatism grew dramatically and eventuated in reaction and the Europe-wide revolution of 1848. A leading role in this attack in the Germanies was taken by materialism. Hopefully, materialism would sap the supernatural sanctions of Christianity and in so doing not only shake the dogma of the churches but also undermine the legitimacy of contemporary absolutist princes who drew strength from purportedly divine rights to power. The reductionist campaign in physiology at the outset may have shared these radical roots but, after the collapse of the revolutions and an uneasy return to traditional theologico-political forms, it was quite evident that German physiological materialism belonged to at least two parties. The first group took aim at religious doctrine, especially the supposed immateriality of the soul, and continued its political agitation. This group (Jacob Moleschott, Ludwig Buchner, and Carl Vogt) concentrated on popularizing science, emphasizing the virtual identity of science and materialism.

The second group, the physiological reductionists, was distinctly more reserved. They were professional scientists, their concerns were overwhelmingly physiological and not political, and they ultimately tended toward political conservatism. Du Bois-Reymond, for example, later became a willing ornament of Prussian nationalism and the new German empire. Three of the leading reductionists—Emil Du Bois-Reymond (1818–1896), Ernst Brücke (1819–1892) and, later, Hermann von Helmholtz (1821–1894)—had been students in the Berlin laboratory of Johannes Müller and were considering the reductionist option by the early 1840s. A fourth and perhaps the most influential member of the group, Carl Ludwig (1816–1895), was trained at Marburg and there acquired the "physicalist tendency" which was the hallmark of the group. Ludwig and the others met in Berlin in 1847, a year before the outbreak of revolution, and there, it is related, cast a plan for a revolution in physiological aspiration and methodology.

Du Bois-Reymond's introduction (1848) to his principal work, *Untersuchungen über thierische Elektrizität*,[3] first vigorously proclaimed the reductionists' ambitions. The enemy was the idea of an autonomous, life-dominating vital force. Support for this doctrine they settled, partly to maintain the force of their polemic, upon the nature-philosophers of the previous generation. The nature-philosophers, they announced, were metaphysicians, not responsible men of science. According to Du Bois-Reymond, the "vital force" was but a "comfortable resting-place where, in Kant's expression, reason finds peace on the cushion of obscure qualities." Physiology must therefore change and change radically. She must

"fulfill her destiny. . . . If one observes the development of our science he cannot fail to note how the vital force daily shrinks to a more confined realm of phenomena, how new areas are increasingly brought under the dominion of physical and chemical forces. . . . [I]t cannot fail that physiology, giving up her special interests, will one day be absorbed into the great unity of the physical sciences; [physiology] will in fact dissolve into organic physics and chemistry."

As exemplification of this new organic physics DuBois-Reymond offered his researches on animal electricity. In the electrical action of the nerves, his realm of special competence, were seen phenomena unusually amenable to the experimental methods of the physical sciences. The early reductionists and their followers were acute and diligent experi-

---

[3] *Researches on animal electricity.*

mentalists. Their instruments, notably Ludwig's kymograph and Du Bois-Reymond's galvanometer and devices for electrical stimulation of tissues, symbolize the high status they assigned to the laboratory.

What had been manifesto in Du Bois-Reymond's introduction soon became systematic dogma. In his influential *Lehrbuch der Physiologie des Menschen* (1852–1856)[4] Ludwig argued that physiology was essentially analysis. It ultimately reduced a living organism to the imponderable fluids (luminiferous aether and electricity) and the chemical elements dependent upon them. No further analysis was possible. Thus, the physiologist must conclude—it was a declaration of faith on Ludwig's part—that all vital phenomena result from attractions and repulsions, that is, from the forces inherent in the irreducible, imponderable fluids and chemical elements of the body. Ludwig's student, Adolf Fick, later restated (1874) the reductionist position with ultimate simplicity and clarity. Physiology, he declared, must take a new direction, a direction that will demonstrate that "vital phenomena" are

"caused by the forces inherent in the material bases of the living organism. Since customarily these forces are divided into chemical and physical we may designate this as the 'chemico-physical' [direction of physiology]. In so far, however, as all forces are in final analysis nothing other than motive forces determined by the interaction of material atoms and in so far as the general science of motion and its causal forces is called mechanics, we must designate the direction of physiological research as truly 'mechanical'."

This declaration offers the heart of the absolute reductionists' credo. Fick, like Ludwig, had elsewhere defended his opinion with the argument that science demands causality. When we introduce the vital force, a "foreign hand" thereby enters the living body and disturbs the normal, autonomous dynamic relationship of the body's atoms. There immediately results a "breakdown of causal connection," and in this disastrous affair "man's need for knowledge" can never acquiesce.

The reductionist thus postulated as the ultimate terms of physiological explanation both force and matter and the indestructable causal bond between every successive event. His was a materialistic and mechanistic universe. The reductionists' terms remain, however, merely premises and are not to be considered necessary conclusions drawn from organic or, indeed, inorganic phenomena. An apparently concrete realm

[4] *Textbook of human physiology.*

of being, that of force, matter, and motion, the domain of mechanics, was prescribed for physiology. The stated intention of the prescription was to replace the idealistic or spiritual categories earlier introduced by nature-philosophers or other nonmechanistic interpreters of life. Both rigid reductionist and systematic nature-philosopher demanded absolute explanation and both were essentially speculative in method and dogmatic in assertion. On this general level, therefore, perhaps the greatest difference between the reductionist and the nature-philosopher lies in the content of their explanatory statements and not in the form or intention of those statements. Herein resides that revolution in ontologies recorded above.

But on a more specific level such comparisons seem less plausible. It is clear that the reductionists were inspired by abhorrence of their predecessors' speculative excesses. No less were they inspired by the successes being daily recorded in physical and chemical research and by the sincere hope that such success might soon be translated to physiology. From physics and chemistry they borrowed methods and instruments and also surely a fair portion of their doctrinaire materialism, and thus much of their own program.

Yet reductionism offered unfulfilled programmatic demands as much as solid accomplishment. Experimental physiology made extraordinary progress during the nineteenth century and the reductionists contributed no less than others to that progress. Theirs was a great if not exclusive role in the definition of fruitful means for the investigation of organic functions. But fully to satisfy their program more than experimental ingenuity was required. Life, they had argued, must be reduced to force and matter and from these first principles all physiological explanation was to begin. This was a goal not to be achieved, for life's complexities were too vast and the explanatory premises distinctly too simple. Despite his early dogmatism, and despite the continued advocacy by some physiologists of a rigorous reductionist position, Du Bois-Reymond came to realize that physiology, like all sciences, faced definite limitations on its explanatory aspirations. Before the "enigma [of] what force and matter are, and how they are to be conceived" he confessed impotence. Now it was necessary to admit "Ignorabimus" before these problems; earlier Du Bois-Reymond had cried "Ignoramus." Very little was left of that assertiveness of 1848 which had dissolved physiology "into organic physics and chemistry." As with the early physiological chemists, the reductionists aimed for comprehensive explanations but had to be satisfied with partial solutions. Their appeal to the modern physiologist and molecular biologist, however, and the pertinence of their claims and experimental practice to recent physiological inquiry requires no emphasis.

Positivism in Physiology

Reductionist and vitalist might fairly agree that their larger objective was one. Their explanatory terms, be they irreducible vital force(s) or the force and matter underlying all natural phenomena, would each satisfactorily answer the great question, What is life? Their answers were their assumptions. It was, however, both possible and highly profitable to pursue serious physiological inquiry without explicitly committing oneself to disputable assertions regarding the essence of life. No figure better exemplifies this sophisticated and self-critical approach to the larger questions of physiological interpretation than Claude Bernard.

Bernard (1813–1878) became the preeminent French physiologist after 1850. His great experimental skill was repeatedly evidenced by major discoveries in the realm of digestion and animal chemistry, pharmacology, and neurophysiology. He was also a deeply concerned observer of the progress and methods of his science and expressed his concern amply in various lecture series (later published) on physiological issues and in tracts devoted to explicit consideration of methodological problems in physiology.

For Bernard only that endeavor which, in a given experimental situation, can repeatedly produce precisely similar results deserves the name of science. Mere observation of organic processes is always interesting but, lacking precisely controlled conditions, evolves only incidental and nonreproducible data. On such data no secure science of physiology could ever be erected. Certainty of the datum, meaning essentially repeatibility of that datum, must become the foundation of physiology and of scientific medicine. Here enters the need for carefully specified experimental procedure. It is absurd, Bernard claimed, to separate observation and experiment, for the latter will necessarily incorporate the former. But experiment meant more, far more, and this additional meaning truly defined the practice of physiology. "Experimentation" Bernard wrote, "is only *provoked observation,* [observation] carried further with the aid of instruments and other means." Experimentation thus necessitated human intervention in the course of vital processes. Physiology was not passive and content simply to amass and coordinate data; physiology must act on and thus control the phenomena it observes and records. The "aim of every science," Bernard wrote, "is to foresee and to act."

Bernard's attention concentrated, therefore, on problems of method. Whatever the laboratory procedures an experimentalist might elect to employ, his business was with phenomena and their determining conditions and not with ultimate causes or the essence of life. "We cannot attain," Bernard declared in 1875,

"to the principle of anything, and the physiologist has nothing more to do with the principle of life than the chemist has with the principle of the affinity of bodies. First causes elude us everywhere, and everywhere alike we can reach only the immediate causes of phenomena. Now, these immediate causes, which are nothing else than the very conditions of phenomena, are capable of as rigorous ascertainment in the sciences of living bodies as in those of lifeless ones."

These words clearly expressed Bernard's positivism. The strict positivist confines himself to phenomena and the ascertainable relations between phenomena. He avoids seeking primary causes and the essence of things. Auguste Comte, creator of the systematic positivist viewpoint, had proclaimed in 1830 that

"only the knowledge of facts is fertile; that the ultimate form for certitude is furnished by the experimental sciences; that the mind, in philosophy as in science, avoids mere verbalism and error only on condition of ceaselessly adhering to experience and renouncing everything which is given *a priori;* that, finally, the domain of 'things in themselves' is inaccessible and our thought may attain only to relations and to laws."

Bernard has been portrayed as the most consistent of scientific positivists and certainly his methodological views, as he so vigorously claimed, are in admirable accord with his actual practice in the physiological laboratory. Bernard never lost from sight that the well-established fact, the postive fact, must form the basis for any generalization in physiological science.

For Bernard, the primary task of the experimentalists was to discover and then to manipulate the "conditions of phenomena." The more varied and carefully designed these manipulations, the greater our knowledge of phenomena and of the all-important interrelations between the phenomena themselves and between phenomena and our instruments of investigation. Bernard's active physiology thus should produce an ever-richer awareness of the relations between functional activities and may do so without concern for the truly unapproachable "principle of life." These conclusions neatly define the activity of Bernard's ideal experimental physiologist. He should emphasize control over all conditions affecting the organism; he should demand reasonable but not excessive precision in his recording of data; he should, in short, master the experimental situation and thus determine "the link between the fact and the idea, between the phenomenon and its conditions." Paul Bert, trained by Bernard and himself a leading experimental physiologist, stated well (1870) the "simple good sense" which he felt characterized French physiology and had stemmed from Bernard:

"To determine the general relationship between phenomena and the experimental conditions of which we are master, to vary these conditions so far as we are able, to arrange and formulate the results so obtained and to attempt to discover, by multiplying our points of view, what kind of modifications our formula must experience in order to be applicable to those complex conditions over which we have no control—such is the true role for us physiologists and such are the limits which [one should not] exceed."

Bernard offered his methodological prescriptions less to define concrete laboratory procedure than to demonstrate the very possibility of a genuine physiological science. The predominant doctrines in Paris during Bernard's student years had been cast by Xavier Bichat and his followers. The Bichatians resolved the body into tissues and to each tissue had assigned a unique combination of properties (see Chapter II). Bernard praised their work and saw in it both a denial of a unitary, omnipotent vital force and a promise that vital phenomena might be rigorously analyzed by attending closely to the special qualities of the body parts. But Bichat had also argued that life truly means spontaneity. The incessant activity of the organism—its most conspicuous and important quality—defied exact measurement and codification into laws. Science—Bernard was thinking of contemporary physics and chemistry —was founded, however, on the existence of stable laws. The regularity of nature was the scientist's necessary presupposition. To bring phenomena into agreement with these regularities opened the possibility of control over the conditions which bring forth desired phenomena. Only here, in a realm where precision and prediction govern, is science possible. By refusing to put life under "any exact law, any constant and settled condition," Bernard believed the vitalistic successors to Bichat had not only asserted that the essence of life is inscrutable but had also argued that no real science of organic function could ever be created.

Bernard, the student of the living organism, wanted neither vitalism nor mechanistic materialism; he wanted physiology as an autonomous science. The nature of the organism demanded this conclusion. The mechanist offered proud claims and precise data but little understanding of the overall or integral behavior of the organism. The extreme vitalist virtually proclaimed such understanding to be unattainable. Bernard's physiological approach was to place the cell at the center of all biological concern. The cell was the common basis of life in plants and animals and its behavior was subject to just those conditions (for example, temperature, nutrient supply, fluid balance, and so on) which the physiologist undertook to define and control. From these thoughts arose Bernard's famous generalization of the internal environment. Every animal, he stated, lives in two environments, the *"external environment* in which the organism lives and an *internal environment* in which the tissue elements [cells] live." The internal environment is the complex fluid (in higher animals, the plasma or liquid part of the blood) which bathes the cells, mediates between the cells and the external world, and is largely responsible for that harmonious integration of myriad diverse physical and chemical events that take place in the organism. "The fixity of the internal environment is," Bernard urged "the [necessary] condition of a free and independent life."

In advancing his notion of the internal environment Bernard never lost sight of broader reflections on physiology and physiological method. The internal environment preserved the integrity of the organism—here was scientific doctrine; it was defined by measurable physical and chemical parameters—here were "conditions" for experimental investigation. The physiologist establishes the relations between vital phenomena. He does not presume to seek the essence of life. Instead of positing a "definition" of life Bernard insisted that the physiologist should gain a "conception" of life, and this meant quite simply taking an *a posteriori* view of vital phenomena. We need, he continued, to "draw a distinction between the metaphysical world and the phenomenal physical world, which serves as its basis, but can borrow nothing from it. . . . We *think* metaphysically, but we *live* and *act* physically."

Bernard and the reductionists agreed that experimentation based on the principles and techniques of physics and chemistry offered the surest grounds for the progress of physiology. Bernard also emphasized the distinctive values of vivisection for physiology, thereby continuing the anatomical and surgical traditions of that science. But Bernard refused to share the metaphysical commitments of his German contemporaries. Whatever else experimentation might teach about life and the organism it could not demonstrate that either could be reduced to force and matter. It could not do so because knowledge of the ultimate nature of vital processes was forever closed to us. At most, force and matter were phenomena or our interpretations based on phenomena; they were not first causes. On similar grounds Bernard's views departed from the metaphysical certainties of dogmatic vitalists.

Physiology

The story of the investigation of respiration illustrates several aspects of the development of physiology during the nineteenth century. That story must not be viewed as a narrative of uninterrupted progress, for this is rarely the case in the development of science and certainly did not occur in respiratory physiology. For the purposes of this brief discussion, however, achievement has been emphasized and attention focused on the investigations culminating in Rubner's comprehensive theory of the 1890s. At the root of respiratory physiology lie the chemical events of combustion and these in turn lead on to physics and the realm of energy. The organism was shown to be a true heat machine. Chemical and physical analyses pointed to the energy sources contained in ingested foodstuffs and suggested possible avenues and means within the body for conversion of that energy into forms useful to the organism. In 1900 the details of these latter operations were still quite obscure and not all physiologists were confident that resolution of these problems was possible. There was confidence, however, that the manifest activities of the organism—locomotion, glandular secretion, nervous transmission, the digestive operations, and numerous others—required, in final analysis, abundant energy for their execution. The century's researches revealed that the requisite energy was indeed available and that the organism's transformation of energy was precisely in accord with the general principles established for energy transformations in the universe at large. For those who cared to do so, here were rich suggestions for redefining the familiar conception of the organism as a machine and for urging the finality of this solution to the question of what is life.

The information upon which these conclusions, be they physiological or metaphysical or simply speculative, were based, was drawn from both observation and experiment. Bernard's remarks require reemphasis: observation, itself a complex term from the viewpoint of methodology, is presupposed by experiment. Experiment, too, is no simple expression. Increasingly during the nineteenth century experiment became the rallying cry of the younger and ambitious physiologists. Their definitions of the term varied but they shared a common objective. Experiment, whatever else it may mean or be, must guarantee control over the appearance and variability of the phenomena under investigation. Whether one proceeds as vivisectionist, relying on surgical intervention in the affairs of the organism, or exploits narrowly the concepts and instruments of the physical sciences, physiologists could agree that mastery of vital phenomena was their achievable goal. These convictions were at the heart of both the experimentalists' practice and Bernard's well-considered reflections on the methods of his science. They depend on a firm belief in the regularity of natural processes and derive as well from an inverse reading of the time-honored conviction that knowledge is power—to control is to know.

CHAPTER VII

The Experimental Ideal

THE FINAL QUARTER of the nineteenth century marks a turning point in the affairs of biology. As biologists focused ever more intently on problems of organic function they transferred their allegiance from the ideal of historical explanation, the critical support for all who had studied organic form and transformation, to the promise extended by the experimental investigation of vital processes. Throughout the century those areas of biology most in the public eye, being also probably those disciplines with the greatest number of trained practitioners, had depended upon historical explanation. Embryology, natural history, evolution theory, even cellular anatomy, were fundamentally historical disciplines. In describing or reconstructing the past the historian of life was persuaded that he possessed in his newly won knowledge not only a fascinating chronicle of events but, more significantly, a deeply satisfying explanation of those events. Wherever it was applied, historical explanation was deemed causal explanation. "The 'historical' conception of nature," the historian John Bagnell Bury wrote, is applicable to celestial and terrestrial beings and has "revolutionized natural science"; this conception "belongs to the same order of thought as the conception of human history as a continuous, genetic, causal process." The biologist-as-historian and the general historian of man and human society dealt with comparable phenomena and assumed necessarily a common mode of explanation.

In tracing the formation of a new cell from a preexisting cell one witnessed the actual process of cell-generation. The same direct confrontation of continuous stages of the developmental or historical process was possible when one examined attentively the complex sequence of events of embryogenesis. In both cases, the observer received a con-

160

crete and seemingly indisputable impression that the daughter cell or a later embryonic stage not only followed in time a related and immediately antecedent cell or stage but was brought into being, was caused, by that same prior cell or stage. The causal argument here employed is clearly *post hoc propter hoc,* and is fully open to the criticism that may always be directed against those who will confuse temporal succession and causal explanation. But such confusion is the heart and very condition of most historical explanation, past and present, and the biologist's use of the argument only ties him more closely to the great intellectual prejudice of nineteenth-century thought.

The evolutionary biologist faced an additional hurdle. The factual record of the history of life on the earth was incomplete. Only in the rarest instances dared he suppose, and that with greatest trepidation, that this particular extinct species gave rise in a determinate number of generations to another particular species. The temporal continuity of living forms was convincing in the broadest sense, but quite deficient when one wished to compare—as one must when seeking a precise and detailed knowledge of the evolutionary process itself—species or organisms forming any two continuous stages in the evolutionary record. Nineteenth-century biologists recognized the problem and attempted to resolve it. Their solution today appears only verbal, but was then regarded as eminently causal. The fact of evolution demanded some connection, preferably a material connection, between all reproducing individuals and between the species which they compose. That connection was designated heredity.

"Evolutionary development has occurred and continues to occur through the agency of material propagation, through ancestral generation and according to the laws of heredity and of the variability and adaptation which modify heredity. All life forms," Ernst Haeckel continued (1866), "even the highest and most complex among them, can arise only by this means—through gradual differentiation and transmutation of the simplest and lowest forms of existence."

The actual nature of the hereditary agency and of the laws, if such there be, governing its action were the greatest mystery to confront early evolutionary biologists. But the need for an hereditary agency, however hypothetical it be and however improbable its various formulations came to appear, is of capital importance to the issue of historical explanation as causal explanation. By means of heredity—minimally defined as the essential (physical) bond between two successively produced individuals—the evolutionary biologist was assured that he might experience that same direct confrontation, so much more accessible to cellular anatomist or embryologist, with the continuous stages of a temporal process. Heredity filled in details in an admittedly deficient historical record and, by so doing, completed the argument for substantial historical explanation of evolutionary events.

Those who employed historical explanation were primarily students of organic form, of the constituent parts (cells) of the body, of the production of the organism by means of these parts and their products (tissues, organs) and of their ultimate disposition in the adult body, of the transformation of one species (a collective entity of similar forms) into another species. Biological interests are, however, considerably wider than this. The study of organic function grew increasingly in importance during the nineteenth century, and herein lie the inspiration, explanatory concepts and laboratory instrumentation for the reorientation of biological thought that marked century's end.

Function displaced form as the goal of biological inquiry. A revolt from morphology set in; a reaction against description and comparison began. Ideals long the valued possessions of physiology—precise, meaning quantitative, delineation of organic phenomena; experimental control over those phenomena; aspiration toward prediction of phenomena— were extended to most and perhaps all domains of biology. It is this vast extension throughout biology's many specialties of the physiologist's ideals and practice that transformed the prospect of biology during the 1880s and 1890s and assured a bright future for the disciplined, ambitious, and unprecedentedly well-trained younger generation of researchers then defining their life's goals. Thomas Henry Huxley had spoken higher truth to his generation and to the vast majority of previous and contemporary biologists when he declared in 1874 that "the parallax of time helps us to the true position of a conception, as the parallax of space helps us to that of a star." Only thirty years later, however, Thomas Hunt Morgan, a student in a direct intellectual succession from Huxley, spoke for a new biology:

"[T]he recognition that only by experimental methods can we hope to place the study of Zoology on a footing with the sciences of chemistry and of physics is a comparatively new conception. . . . I think it will be generally admitted that at the present time there is greater need for experimental work than for descriptive and observational study."

Morgan expressed no desire to disparage the work of the still sizable corps of able biologists dedicated to the "descriptive and historical

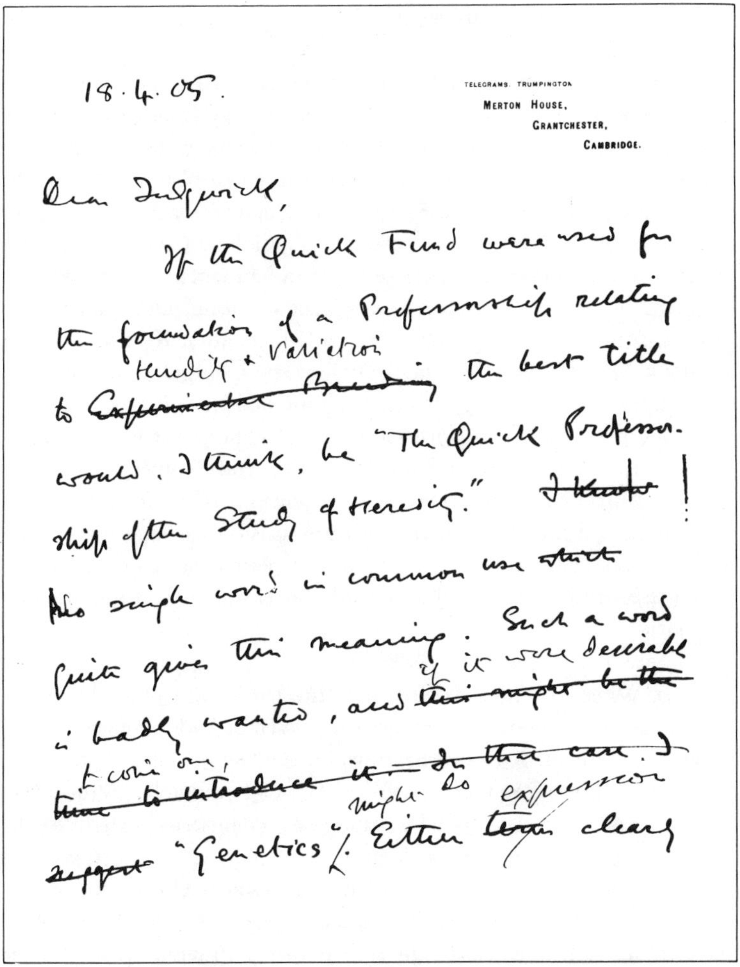

18.4.05.

TELEGRAMS: TRUMPINGTON
MERTON HOUSE,
GRANTCHESTER,
CAMBRIDGE.

Dear Sedgwick,

If the Quick Fund were used for the foundation of a Professorship relating to ~~Experimental Breeding~~ Heredity + Variation the best title would, I think, be "The Quick Professorship of the Study of Heredity." ~~I know~~ No single word in common use ~~strictly~~ quite gives this meaning. Such a word is badly wanted, and ~~this might be the~~ if it were desirable ~~time to introduce it. In that case I~~ to coin one, ~~suggest~~ "Genetics" might do. Either ~~term~~ expression clearly

Figure 7.1 A bequest was made to Cambridge University for the establishment of a new professorial chair in biology. The chair ultimately was assigned to protozoology, but that decision was made only after William Bateson, the leading English advocate of Mendelian studies of heredity, had appealed to his Cambridge master, Professor Adam Sedgwick, to help direct the new fund toward the promising domain of genetics. Bateson here coins the term Genetics; it first appeared in print in 1906: "If the Quick Fund were used for the foundation of a Professorship relating to Heredity and Variation the best title would, I think, be 'The Quick Professorship of the study of Heredity.' No single word in common use quite gives this meaning. Such a word is badly wanted, and if it were desirable to coin one, 'Genetics' might do. Either expression clearly {includes Variation and the cognate phenomena.}." (William Bateson, 1905: Gregory Bateson.)

problems of biology." His widely shared objective was not to destroy but to build.

The advances made in understanding the essential respiratory process (the liberation from foodstuffs of heat and energy to power the living organism) exemplifies nineteenth-century progress in physiology's experimental instrumentation and conceptual sophistication. Much the same point could be made stressing the development of neurophysiology or aspects of cardiovascular physiology. Physiology's success seemed a function of her impassioned concern for experimental inquiry. Experiment meant, above all, control of pertinent environmental conditions— from general control of temperature or illumination to, for example, the more specific level of stimulation by specific agents of particular body parts—and included as well the provision of suitable means for recording precisely the organism's responses. Experiment was an instrument of discovery, a method for verification and frequently the basis of practical laboratory instruction. Given human mastery of physiological conditions, and the always necessary reaffirmation that nature is regular and not capricious in her behavior, the experimentalist could vary at will these conditions and thus record the ensuing spectrum of vital responses.

The great attraction of experimental analysis of organic functions is obvious. When carefully executed and directed toward properly formulated questions experiment promised to return reliable, indeed certain, knowledge. For many this certainty was a function of one's instruments and these, while recognizably not perfect, were infinitely preferable to the apparent subjectivity of observation and comparison as effected by the traditional biologist. Physiological instruments promised, moreover, to give access to various aspects of the organism heretofore ignored or effectively closed to unaided inspection. Even vivisection, while short on complex instrumentation, in skilled hands brought forth reliable data and proved once again its central role in physiological investigation.

No biologist could long resist the temptation of such a promising method. The outstanding instance of redefining the direction of a biological discipline—embryology—was given by Wilhelm His and especially Wilhelm Roux (see Chapter III). Their interest in the possibility and means for the manipulation of the conditions of embryonic development largely created the still-vital field of experimental embryology. His saw no value in simply rejecting the observational and comparative approach that had stemmed from the early epigeneticists and notably von Baer. He recognized, however, the incompleteness of their work and in consequence argued that "the general scientific methods of measuring, of weighing and of determining volumes cannot be neglected in embryological work."

Similar appeals were launched by other voices. Young men trained during the 1880s in the varieties of Darwinism (see Chapter IV) soon expressed dissatisfaction with the principal objective set before them, the reconstruction by means of comparative anatomy, embryology, and paleontology of the evolutionary history of plants and animals. It was now much more important, they held, to attempt to analyze the process of change than to seek probably unattainable detail regarding the products of change. The only method truly suited to this new goal was experimental analysis of plant and animal breeding behavior and of the disposition among the offspring of heritable traits. Such work promised to answer the outstanding questions of evolutionary theory. "In experimental transformism [basically, a study of hybridization and environmental influences]," wrote Henry de Varigny in his aptly titled *Experimental Evolution* (1892), "lies the only test which we can apply to the evolutionary theory." Mendel's discoveries, however, soon became known and the experimental investigation of heredity and variation immediately became the core of the new science of genetics. Here again, and now in the very darkest heartland of nineteenth-century evolutionary speculation, experiment appeared and it brought clarity and (statistical) precision. The new direction is forcefully portrayed in the simple title of Wilhelm Johannsen's great work, *Elemente der exacten Erblichkeitslehre* (1909),[1] the earliest major compendium of knowledge in genetics.

The experimental ideal appeared elsewhere. After 1870 the study of the cell was directed increasingly toward the physical conditions, principally membrane behavior and osmotic phenomena, which ensured that structure's chemical operations. The fertilization process was artificially induced by chemical and physical stimulants. Cultures of isolated tissues were prepared and perpetuated. Here were powerful techniques that opened a prospect for unlimited experimental manipulation and assessment of the critical processes of reproduction and growth. The brilliant laboratory operations and experimental cautions of Robert Koch, with Louis Pasteur the leading personage in the creation of scientific bacteriology and the germ theory of disease, virtually ensured by 1880

---

[1] *Elements of the exact theory of heredity.*

that both medicine and preventive health measures would require a new foundation to be won only in the laboratory. Even conscious states of mind and instincts were to be pursued by the experimentalist; beginning in the 1870s the laboratory became a training ground for psychologists. In place of customary recourse to introspection and description of mental states the new physiological psychologists sought by means of exact instruments and suitably applied stimuli to create a rigorous science of psychology.

The historical ideal was not lost to biology. Rather, it was supplemented, but supplemented so forcefully as to be placed in a distinctly lesser position. In the book of biology in 1900 was written a persuasive word: experiment. The ranks of the senior generation, the generation of the great evolutionists, were rapidly thinning and new recruits to biology were inclined to be vocal partisans of a physiological view of biology's concerns. Function, now studied increasingly on the physical and chemical level and with the aid of the conceptual and laboratory implements of those sciences, meant vital process, the day-to-day, second-to-second events whose sum total was life. Experimental physiology had established a model approach to such events and that approach—experiment—offered little encouragement to the traditional historian of life. But to others, to the embryologist, the bacteriologist, or the student of heredity and variation, the model was truly the new ideal. In its name—experiment—was set in motion a campaign to revolutionize the goals and methods of biology.

Bibliography

BIOLOGY

General histories of biology, especially when dealing with the nineteenth century, rarely offer more than a plain record of names, dates, and discoveries. An exception, however, is Erik Nordenskiøld's *History of biology* (New York, 1928). Nordenskiøld's coverage of anatomy, embryology, and natural history is more reliable and complete than his account of physiology and evolution. Another exception, more limited in scope but still today exceedingly useful, is Julius von Sachs, *History of botany (1530–1860)*, trans. H. E. F. Garnsey (Oxford, 1890). The only work in English devoted exclusively to nineteenth-century biology is J. Arthur Thomson, *The science of life. An outline of the history of biology and its recent advances* (Chicago, 1899). Exceptionally valuable discussions of morphology, historical explanation in geology and biology, and the general problems of life are provided by John T. Merz in his classic *A history of European thought in the nineteenth century* (reprint: New York, 1964), II, 200–464. The development of many aspects of biology is related to the progress of medicine by R. H. Shryock, *The development of modern medicine* (New York, 1947).

Biology and her subdisciplines are included by Everett Mendelsohn in a review of problems of scientific professionalization: "The emergence of science as a profession in nineteenth-century Europe," *The management of scientists,* ed. K. Hill (Boston, 1964). See also the excellent analysis by Joseph Ben-David, "Scientific productivity and academic organization in nineteenth-century medicine," *American Sociological Review* 25 (1960): 828–843. A sweeping and polemical account of the patronage of biological study and institutions, particularly

167

museums, is given by William Swainson, *A preliminary discourse on the study of natural history* (London, 1839), 296–450. On the rise of the physiological institutes see Stephen d'Irsay, *Histoire des universités françaises et étrangères* (Paris, 1935), II, 280–290.

Two essays deal generally with terminological and other problems regarding the use of historical explanation in biology (see also Merz, above): Walter Baron and Bernhard Sticker, "Ansätze zur historischen Denkweise in der Naturforschung an der Wende vom 18. zum 19. Jahrhundert," *Sudhoffs Archiv* 47 (1963): 19–35; Walter Baron, "Wissenschaftsgeschichtliche Analyse der Begriffe Entwicklung, Abstammung und Entstehung im 19. Jahrhundert," *Tecknikgeschichte* 35 (1968): 68–79.

FORM

No work has replaced E. S. Russell, *Form and function* (London, 1916), as a comprehensive if often technically uncompromising account of nineteenth-century views on the nature and formation of the organism. Russell's book largely defined its field and by so doing made intelligible a vast amount of otherwise confusing and seemingly unrelated biological study. Texts illustrating differing viewpoints on the proper study of organic form, together with a brief historical essay and bibliographical note, have been published by William Coleman, *The interpretation of animal form* (New York, 1967). *The evolution of the microscope* by S. Bradbury (Oxford, 1967) supplants previous studies of this instrument.

Marc Klein's pioneering *Histoire des origines de la théorie cellulaire* (Paris, 1936) should be supplemented with Arthur Hughes, *A history of cytology* (London, 1959), a readable and dependable essay, and the five-part exacting analysis by John R. Baker, "The cell-theory: A restatement, history and critique," *Quarterly Review of Microscopical Science* 89 (1949): 103–125; 90 (1949): 87–108; 93 (1952): 157–190; 94 (1953): 407–440; 96 (1955): 449–481. Thomas S. Hall devotes a major section of his *Ideas of life and matter* (Chicago, 1969), II, 121–304, to "Tissue, cell and molecule, 1800–1860." A stimulating essay, relating the creation and development of cell theory to the larger issue of biological individuality, is Georges Canguilhem, "La théorie cellulaire," *La connaissance de la vie,* ed. 2 (Paris, 1967), 43–80.

On the founders and advocates of cell doctrine see Rembert Watermann, *Theodor Schwann, Leben und Werk* (Düsseldorf, 1960); Marcel Florkin, *Naissance et déviation de la théorie cellulaire dans l'oeuvre de Theodore Schwann* (Paris, 1960) and Erwin Ackerknecht, *Rudolf Virchow. Doctor, statesman, anthropologist* (Madison, 1953). A collection of Virchow's essays, translated into English, was published by Leland J. Rather, *Disease, life and man* (Stanford, 1958). Mathias Jacob Schleiden has yet to attract a modern biographer; see, however, the valuable introduction by Jacob Lorch to a new edition of Schleiden's great work, *Principles of scientific botany* (reprint: New York, 1969), ix–xxxiv.

For the idea *anatomia animata* or functional anatomy see Stephen d'Irsay, *Albrecht von Haller. Eine Studie zur Geistesgeschichte der Aufklärung* (Leipzig, 1930), 34–45; also Martin J. S. Rudwick, "The inference of function from structure in fossils," *British Journal for the Philosophy of Science* 15 (1964): 27–40. On Bichat and his conception of tissues the basic study remains A. Arène, "Essai sur la philosophie de Xavier Bichat," *Archives d'Anthropologie Criminelle, de Médecine Légale et de Psychologie Normale et Pathologique* 25 (1911): 791–825.

The best general introduction to the problems and personalities of nineteenth-century embryology is Jane M. Oppenheimer, *Essays in the history of embryology and biology* (Cambridge, Mass., 1967). Elizabeth Gasking, *Investigations into generation. 1651–1828* (Baltimore, 1966) provides a careful review and interesting interpretation of embryology to ca. 1850. One should still consult A. W. Meyer, *The rise of embryology* (Stanford, 1939). Two essays by Frederick B. Churchill open to view initial steps and methodological dilemmas in the creation during the late nineteenth century of experimental embryology: "August Weismann and a break from tradition," *Journal of the History of Biology* 1 (1968): 91–112; "From machine-theory to entelechy: Two studies in developmental teleology," *Ibid.* 2 (1969): 165–185. On the master of comparative embryology, von Baer, see Boris Raikov, *Karl Ernst von Baer 1792–1876. Sein Leben und sein Werk* (Leipzig, 1968).

A simple record of the development of German *Naturphilosophie* and a full assessment of its role in nineteenth-century biology is yet to be made. Stephen F. Mason, *Main currents of scientific thought* (New York, 1956), 280–290 provides a brief but suggestive overview. Owsei Temkin touches on related themes but particularly on the possible analogy between the life cycle of man and the history of life in "German concepts of ontogeny and history around 1800," *Bulletin of the*

History of Medicine **24** (1950): 227–246. M. H. Abrams' *The mirror and the lamp. Romantic theory and the literary tradition* (New York, 1953), 156–225, is a work of outstanding interest to students of nineteenth-century biology; Abrams explores the meaning of "mechanical" and "organic" theories of poetic invention and relates these to the metaphysical tradition which doubtlessly also underlies the biologists' use of *Naturphilosophie*. See also Brigitte Hoppe, "Deutscher Idealismus und Naturforschung. Werdegang und Werk von Alexander Braun (1805 bis 1877)," *Technikgeschichte* **36** (1969): 111–132.

R. C. Olby has described nineteenth-century studies of heredity and variation through the first decade of Mendelism: *Origins of Mendelism* (London, 1966). Similar material is covered by Hans Stubbe, *Kurze Geschichte der Genetik bis zur Wiederentdeckung der Vererbungsregeln Gregor Mendels,* ed. 2 (Jena, 1965); Stubbe provides an ample bibliography. Gerald L. Geison ("The Protoplasmic theory of life and the vitalist-mechanist debate," *Isis* **60** [1969]: 273–292) thoroughly explores one of the bolder extrapolations of the early cell theory. Cytological work establishing the cell nucleus and probably the chromosomes as the vehicle of inheritance is described by William Coleman, "Cell, nucleus and inheritance: an historical study," *Proceedings of the American Philosophical Society* **109** (1965): 124–158. Edmund B. Wilson, a leading American cytologist, was perhaps the best witness to the century's activity in the study of the cell and of individual development. His classic monograph, *The cell in development and inheritance* (1896), reviews the state-of-the-science in the last decade before the rediscovery of Mendel. This work has been reprinted (New York, 1966) with an important introduction by H. J. Muller.

TRANSFORMATION

The literature on the history of evolutionary theory and on Charles Darwin is enormous. The quality and originality of much of this publication is, however, slight and the reader desperately needs a trustworthy guide in the wilderness. The centennial year of Darwin's *Origin,* 1959, was the cause of a fresh spate of publication. A useful guide to studies published through 1963 is provided by Bert J. Loewenberg, "Darwin and Darwin studies, 1959–1963," *History of Science* **4** (1965): 15–64. While Loewenberg's treatment of this literature is, regrettably, generally uncritical, his bibliography is quite full; it also offers indirect access to items issued before 1959.

A major achievement of the centennial period is *Darwin's century. Evolution and the men who discovered it* by Loren Eiseley (Garden City, 1958). Eiseley's account of events through 1859 is comprehensive and highly readable; his report on the post-Darwinian period is episodic and creates no satisfying picture. Indeed, scholarly study of the development of biological interests after 1859 has itself been desultory and remains an area demanding systematic exploration. John C. Greene, in *The death of Adam. Evolution and its impact on Western thought* (Ames, 1959), covers considerably more than the nineteenth century. His book is valuable for its constant attention to the presumed evolutionary history of man and also for its splendid illustrations. An important collection of often highly original essays on the history of evolution is H. Bentley Glass et al., *Forerunners of Darwin, 1745–1859* (reprint: Baltimore, 1968). Maurice Mandelbaum, in "The scientific background of evolutionary theory in biology," *Journal of the History of Ideas* **18** (1957): 342–361, examines biological and religious elements in the casting of evolutionary thought; see also his "Darwin's religious views," Ibid. **19** (1958): 363–378. David L. Hull, "The metaphysics of evolution." *British Journal for the History of Science* **3** (1967): 309–377, incisively reviews arguments from antiquity through the nineteenth century regarding the possible "reality" of species.

No truly satisfactory biography of Darwin the biologist exists. Paul B. Sears, *Charles Darwin. The naturalist as a cultural force* (New York, 1950) and Gavin de Beer, *Charles Darwin* (London, 1963) are brief and reliable accounts. The most recent major study is Gertrude Himmelfarb, *Darwin and the Darwinian revolution* (London, 1959); great caution is needed in accepting Himmelfarb's presentation of Darwin's scientific work. Geoffrey West's *Charles Darwin. A portrait* (New Haven, 1938) has not been superseded as a sensitive character portrait. For the first definition of Darwin's views, see Sydney Smith, "The origin of 'The Origin,' " *The Advancement of Science* No. 64 (1960): 391–401.

The scientific activities of Alfred Russell Wallace are related by Wilma George, *Biologist philosopher. A study of the life and writings of Alfred Russell Wallace* (London, 1964). One should also consult H. Lewis McKinney's important article, "Alfred Russell Wallace and the discovery of natural selection," *Journal of the History of Medicine* **21**

(1966): 333–357. G. Beddall has reexamined the apparent ambiguities in the relationship between Wallace and Darwin and emphasizes the two naturalists' studies in biogeography: "Wallace, Darwin and natural selection. A study in the development of ideas and attitudes," *Journal of the History of Biology* 1 (1968): 261–323. Camille Limoges, *La sélection naturelle. Étude sur la première constitution d'un concept (1837–1859)* (Paris, 1970), discusses amply (using the Darwin manuscripts) the development of the idea of natural selection.

The scope of Darwin's publications is exhibited by *The works of Charles Darwin. An annotated bibliographical hand list* prepared by R. B. Freeman (London, 1965). The first edition (1859) of Darwin's principal work has been reissued in facsimile and with an introduction by Ernst Mayr: *On the origin of species* (Cambridge, Mass., 1964). *The autobiography of Charles Darwin, 1809–1882,* a personal document of great interest and Darwin's most amiable essay, has been published in unexpurgated form by Nora Barlow (London, 1958). Darwin, and indeed most Victorian naturalists, were at their best in private correspondence, private when written but doubtlessly often penned with an eye on eventual publication. Nora Barlow has published Darwin's correspondence with his beloved mentor, Henslow—a collection of letters revealing the development of Darwin's evolutionary ideas: *Darwin and Henslow. The growth of an idea. Letters 1831–1860* (London, 1967).

Loewenberg lists several of the biographical and other studies dealing with the many nineteenth-century naturalists who from one viewpoint or another confronted the "species problem." Remarkably enough, there is not yet a full, analytic account of Lamarck's career and thought; see, however, Marcel Landrieu, *Lamarck. Le fondateur du transformisme* (Paris, 1909); Charles C. Gillispie, "The formation of Lamarck's evolutionary theory," *Archives Intérnationales d'Histoire des Sciences* 9 (1956): 323–338. On Cuvier, see William Coleman, *Georges Cuvier, zoologist. A study in the history of evolution theory* (Cambridge, Mass., 1964), also, "Journées d'Études: George Cuvier," *Revue d'Histoire des Sciences* 23 (1970): 1–92; on Agassiz: Edward Lurie, *Louis Agassiz. A life in science* (Chicago, 1960), also Ernst Mayr, "Agassiz, Darwin and evolution," *Harvard Library Bulletin* 13 (1959): 165–194; on Asa Gray, Darwin's foremost American supporter: A. Hunter Dupree, *Asa Gray. 1810–1888* (Cambridge, Mass., 1959); on Hooker, Darwin's leading British supporter: W. B. Turrill, *Joseph Dalton Hooker* (London, 1963). Thomas Henry Huxley and Richard Owen, outspoken protagonists of opposing positions with regard to Darwin's hypothesis of natural selection, have fared no better than Lamarck. P. Chalmers Mitchell, *Thomas Henry Huxley. A sketch of his life and work* (New York, 1900) is probably the best albeit conventional account of Huxley's activities. On Owen there is nothing of value, not even a useful *Life and letters* so characteristic of Victorian times; the Owen volumes offer a particularly pious and dreadful example of the *genre*. Studies on evolutionary biologists and their critics active outside of Britain and the United States are scandalously wanting; most notable in this regard is Ernst Haeckel. See, however, the excellent review by Piet Smit, "Ernst Haeckel and his Generelle Morphologie: an evaluation," *Janus* 54 (1967): 236–252. Also, Ernst Gaupp, *August Weismann, Sein Leben und sein Werk* (Jena, 1917), and the excellent study by Fritz Baltzer, *Theodor Boveri. Life and work of a great biologist,* trans. D. Rudnick (Berkeley, 1967). Georg Uschmann's report on zoological instruction and research at the university in Jena, the center for Haeckel's evolutionary propaganda, provides a well-documented portrayal of the institutional basis of German academic biology: *Geschichte der Zoologie und der zoologischen Anstalten in Jena, 1779–1919* (Jena, 1959). A delightful and very personal account of scientific doings in the German universities is given by Richard Goldschmidt, *Portraits from memory. Recollections of a zoologist* (Seattle, 1956).

Archibald Geike's *The founders of geology,* ed. 2 (London, 1905) is still of value; see also the short history of geology by C. C. Beringer: *Geschichte der Geologie und der geologischen Weltbildes* (Stuttgart, 1954). *Charles Lyell* by Edward Bailey (London, 1962), slight as it is, is the only recent biography of the geologist (a multi-volume life is in preparation by Leonard G. Wilson). On geological forces, see Reijer Hooykaas, *Natural law and divine miracle. The principle of uniformity in geology, biology and theology,* ed. 2 (Leiden, 1963); Leonard G. Wilson, "The origins of Charles Lyell's uniformitarianism," *Special Paper no. 89: Geological Society of America* 1967: 35–62; Martin J. Rudwick, "Lyell on Etna, and the antiquity of the earth," Cecil J. Schneer, ed. *Toward a history of geology* (Cambridge, Mass., 1969), 288–304; also Rudwick's "The strategy of Lyell's *Principles of geology*", *Isis 61* (1970): 4–33. A fundamental contribution both to our knowledge of Lyell and of the development of evolution theory is Leonard G. Wilson, ed. *Sir Charles Lyell's scientific journals on the species question* (New Haven, 1970). "Time" as a factor in early geological understanding is the subject of Francis Haber, *The age of the world. Moses to Darwin* (Baltimore, 1959).

Evolutionary biology after Darwin is, as noted above, little explored. A convenient but nonrigorous review of various evolutionary hypotheses, including Darwin's, is offered by Yves Delage and Marie Goldsmith, *The theories of evolution,* trans. A. Triden (New York, 1912). The altogether remarkable book of Vernon L. Kellogg, *Darwinism today* (New York, 1907), provides a stimulating assessment of the utter confusion into which evolutionary doctrine had plunged by century's end. Two essays by Garland E. Allen examine closely related aspects of the evolutionists' uncertainties: "Thomas Hunt Morgan and the problem of natural selection," *Journal of the History of Biology* 1 (1968): 113–139; "Hugo de Vries and the reception of the Mutation Theory," Ibid. 2 (1969): 55–88. Four unpublished doctoral dissertations treat the scientific reception of Darwin's evolutionary theory in various nations: Pierce C. Mullen, *The preconditions and reception of Darwinian biology in Germany, 1800–1870* (University of California, Berkeley, 1964); Robert E. Stebbins, *French reactions to Darwin, 1859–1882* (University of Minnesota, 1965); Edward J. Pfeiffer, *The reception of Darwinism in the United States* (Brown University, 1965); and Joe D. Burchfield, *The age of the earth. The theories and influence of Lord Kelvin* (Johns Hopkins University, 1969).

Alvar Ellegård has minutely assessed the British public reaction to Darwin: *Darwin and the general reader. The reception of Darwin's theory of evolution in the British periodical press, 1859–1872* (Göteborg, 1958); see also Ellegård's essay "The Darwinian theory and nineteenth-century philosophies of science," *Journal of the History of Ideas* 18 (1957): 362–393. For Darwinian influences in America see Stow Persons, ed., *Evolutionary thought in America* (New Haven, 1950); George H. Daniels, *Darwinism comes to America* (Waltham, 1968). Richard Hofstadter, *Social Darwinism in American thought,* ed. 2 (Boston, 1955) gives unparalleled coverage of its subject. In *Rendezvous with destiny,* ed. 2 (New York, 1956) Eric F. Goldman examines the possible reformist aspect of American social Darwinism. A work less comprehensive than Hofstadter but no less valuable, especially as it deals with the heretofore poorly investigated German ramifications of the social application of Darwinian precepts, is Hedwig Conrad-Martius, *Utopien des Menschenzüchtung; der Sozialdarwinismus und seine Folgen (Munich, 1955).* Of outstanding importance regarding the interaction of the sciences and other trends in thought, both before and after Darwin, is Robert M. Young, "Malthus and the evolutionists: the common context of biological and social theory," *Past*

and Present No. 43 (1969): 109–141. The theological explosion supposedly precipitated by Darwin is the best-known but by no means
most fully studied facet of the coming of evolutionary theory. The
background to the dispute, as well as a superb bibliographical introduction to the entire subject, is amply presented by Charles C. Gillispie,
*Genesis and geology. A study in the relations of scientific thought,
natural theology, and social opinion in Great Britain, 1790–1850.*
(Cambridge, Mass., 1951). See also Ernst Benz, *Evolution and Christian hope,* trans. H. G. Frank (Garden City, 1966); John Dillenberger,
Protestant thought and natural science (Garden City, 1960). The
Anglican reaction to Darwin is briefly described by John Kent in his
pamphlet, *From Darwin to Blatchford. The role of Darwinism in
Christian apologetic, 1875–1910* (London, 1966). The whole subject
requires a fresh appraisal and the views of Catholic commentators
deserve particular attention.

MAN

The history of the study of man touches perilously near every aspect of
individual and social human endeavor. Interests strictly anthropological
in nature are uncommonly difficult to circumscribe but one useful approach to the problem is historical analysis. Three recent works present
valuable assessments of anthropological doctrine during the nineteenth,
its formative century. *Evolution and society. A study in Victorian social
theory* (Cambridge, 1966) by J. W. Burrow is a difficult but immensely
rewarding study of the decisive years, 1860–1890, in Britain. Paul
Mercier, *Histoire de l'anthropologie* (Paris, 1966), emphasizes the
modern period and is particularly useful for its discussion of the rise of
social anthropology. A comprehensive review of all aspects of the development of anthropology plus an invaluable bibliographical guide to
literature on the subject is provided by Wilhelm E. Mühlmann,
Geschichte der Anthropologie, ed. 2 (Frankfurt am Main, 1968).
George W. Stocking has collected his many stimulating articles on the
history of anthropology and the elaboration of its methods in *Race,
culture and evolution: essays in the history of anthropology* (New
York, 1968). More popular works are H. R. Hayes, *From ape to angel.
An informal history of social anthropology* (New York, 1958); Abram
Kardiner and Edward Preble, *They studied man* (New York, 1961).
Older studies still of considerable utility include Robert H. Lowie, *The*

history of ethnological theory (New York, 1937) and T. K. Penniman, *A hundred years of anthropology* (London, 1935).

The distinction between anthropology and sociology is easily lost in the early years of these disciplines; histories of the latter add much to understanding the development of the former. See particularly Howard Becker and Harry E. Barnes, *Social thought from lore to science,* ed. 2 (Washington, 1952), an immense work literally bursting with information; Emory S. Bogardus, *The development of social thought* (New York, 1940); and Robert Mackintosh, *From Comte to Benjamin Kidd. The appeal to biology or evolution for human guidance* (New York, 1899), a work especially interesting for its author's contemporary and critical appraisal. The background to American sociological emphasis on the stability or homeostasis of the social organism is neatly examined by Cynthia E. Russett, *The concept of equilibrium in American social thought* (New Haven, 1966). E. E. Evans-Pritchard presents an excellent general account of the beginnings and concerns of social anthropology in *Social anthropology and other essays* (reprint: New York, 1964), 1–134.

A fascinating account of anthropology's gestation period is given by Margaret T. Hodgen, *Early anthropology in the sixteenth and seventeenth centuries* (Philadelphia, 1964). Western man's contact with an alien culture stands bared in Roy H. Pearce, *Savagism and civilization. A study of the Indian and the American mind,* ed. 2 (Baltimore, 1965). Two collections of texts offer the best introduction to the interests and philosophico-moral basis of eighteenth-century thought regarding man in his varied conditions: J. S. Slotkin, ed. *Readings in early anthropology* (Chicago, 1965); Louis Schneider, *The Scottish moralists. On human nature and society* (Chicago, 1967). Gladys Bryson, *Man and society* (Princeton, 1945), thoroughly reviews the doctrines of the Scottish moralists.

Much has been written, often of a disputacious nature, on the "comparative method." See especially Erwin H. Ackerknecht, "On the comparative method in anthropology," R. F. Spencer, ed., *Method and perspective in anthropology. Papers in honor of Wilson D. Wallis* (Minneapolis, 1954), 117–125; Kenneth E. Bock, "The comparative method of anthropology," *Comparative Studies in Society and History* 2 (1966): 269–280; and William N. Fenton, "J.-F. Lafitau (1681–1746), precursor of scientific anthropology," *Southwestern Journal of Anthropology* 25 (1969): 173–187.

Kenneth Bock, "Darwin and social theory," *Philosophy of Science*

22 (1956): 123–134, discusses the supposedly intimate relationship between biological and social evolutionary doctrines. More abundant information on this topic is provided by Idus L. Murphree, "The evolutionary anthropologists: the progress of mankind. The concepts of progress and culture in the thought of John Lubbock, Edward B. Tylor and Lewis H. Morgan," *Proceedings of the American Philosophical Society* 105 (1961): 265–300.

Glyn Daniel, *The idea of prehistory* (Cleveland, 1963) is a pleasurable introduction to the history of archeology. Important archeological texts are presented by Robert F. Heizer, ed., *Man's discovery of his past. Literary landmarks in archeology* (Englewood Cliffs, 1962). The recovery of the remains of modern man is described by John R. Baker, "Cro-magnon man, 1868–1968," *Endeavor* 27 (1968): 87–90.

A. Irving Hallowell's, "The beginnings of anthropology in America," Frederica de Laguna, ed., *Selected papers from the American anthropologist, 1888–1920* (Evanston, 1960), 1–104, covers a much broader range of material than its title suggests. So, too, does Jacob W. Gruber, "Horatio Hale and the development of American anthropology," *Proceedings of the American Philosophical Society* 111 (1967): 5–37. On racism in American thought, see the major review and revision of early opinion by Winthrop D. Jordan, *White over black. American attitudes toward the negro, 1550–1812* (Chapel Hill, 1968); William Stanton's full narrative of the "American school" of anthropology, *The leopard's spots. Scientific attitudes toward race in America, 1815–1859* (Chicago, 1960); T. F. Gossett, *Race: the history of an idea in America* (Dallas, 1963); and Mark H. Haller's outstanding survey, *Eugenics. Hereditarian attitudes in American thought.* (New Brunswick, 1963).

Biographical and analytic accounts, of markedly varying quality, of the early advocates and later proponents of the possibility and procedures of the social sciences are: Frank N. Manual, *The prophets of Paris* (Cambridge, Mass., 1960) which includes an excellent discussion of the background and development of Comte's social views; Keith N. Baker, "Scientism, elitism and liberalism: the case of Condorcet," *Studies on Voltaire and the Eighteenth Century* 55 (1967): 129–165; Carl Resek, *Lewis Henry Morgan. American scholar* (Chicago, 1960); Emile Durkheim et al., *Essays on sociology and philosophy, with appraisals of his life and thought,* ed. K. H. Wolff (reprint: New York, 1964); Melville J. Herskovits, *Franz Boas. The science of man in the making* (New York, 1953); and Raymond Firth, ed., *Man and culture. An evaluation of the work of Malinowski* (London, 1957). There exists no satisfying full account of England's most eager metaphysician but see Hector Macpherson, *Herbert Spencer. The man and his work,* ed. 2 (London, 1901); J. Arthur Thomson, *Herbert Spencer* (London, 1906), a work especially valuable for its account of Spencer's biological opinions. Sydney Eisen compares Spencer's and Comte's views on the important matter of the classification of the sciences in "Herbert Spencer and the spectre of Comte," *Journal of British Studies* 7 (1967): 48–67.

FUNCTION

Of the grand domain of nineteenth-century physiology there exists in English no comprehensive and reliable survey; even works whose value is slight are rare. A general introduction to the problems and schools of thought of the period is K. E. Rothschuh, *Geschichte der Physiologie* (Berlin, 1953), 91–224, a most useful work. Rothschuh reemphasizes the broader themes of his subject in "Ursprünge und Wandlungen der physiologischen Denkweise im 19. Jahrhundert," *Physiologie im Werden* (Stuttgart, 1969), 155–181. Georges Canguilhem's exceptional essay, "La constitution de la physiologie comme science," *Études d'histoire et de philosophie des sciences* (Paris, 1968), 226–273, examines the convergence of disciplines and interests which created a science. Recent literature dealing with physiological matters, particularly of the earlier part of the century, is reviewed and listed by Everett Mendelsohn, "The biological sciences in the nineteenth century: some problems and sources," *History of Science* 3 (1964): 39–59. *Selected readings in the history of physiology* by John F. Fulton and Leonard G. Wilson, ed. 2 (Springfield, Ill., 1966) offers commentary and many texts from the nineteenth century, with a clear emphasis on medical physiology. A provocative discussion of the place of physiology in the training and work of the physician is presented by Owsei Temkin, "The dependence of medicine on basic scientific thought," C. McC. Brooks and P. F. Cranefield, eds., *The historical development of physiological thought* (New York, 1959), 5–21. A highly suggestive record of the introduction and utilization of physiological instruments is given by Dietmar Rapp, *Die Entwicklung der physiologischen Methodik von 1784 bis 1911. Eine quantitative Untersuchung* (Münster, 1970).

The background to nineteenth-century respiratory physiology is amply related by Everett Mendelsohn, *Heat and life. The development of the theory of animal heat* (Cambridge, Mass., 1964). G. J. Goodfield provides a more intensive examination of this subject for the years 1800–1850, together with an attempt to resolve the mechanist-vitalist controversy: *The growth of scientific physiology* (London, 1960). The principal study of nineteenth-century respiratory physiology is the unpublished doctoral dissertation of Charles A. Culotta: *A history of respiratory theory: Lavoisier to Paul Bert, 1777–1880* (University of Wisconsin, 1968): see also Culotta's "Tissue oxidation and theoretical physiology: Bernard, Ludwig and Pflüger," *Bulletin of the History of Medicine* 44 (1970): 109–140 and Bradley T. Scheer, "The development of the concept of tissue respiration," *Annals of Science* 4 (1939): 295–305. David Keilin's *History of cell respiration and cytochrome* (Cambridge, 1966) concentrates on the twentieth century; his discussion (1–139) of earlier investigations is, nevertheless, of capital importance. Thomas S. Kuhn, "Energy conservation as an example of simultaneous discovery," *Critical problems in the history of science,* ed. M. Clagett (Madison, 1959), 321–356, is a fundamental review of the topic. The reader will find abundant discussion of the nature and implications of the energy doctrine in William R. Grove et al., *The correlation and conservation of forces. A series of expositions,* ed. 6 (London, 1874). Most useful in this regard is George Rosen, "The conservation of energy and the study of metabolism," Brooks and Cranefield, eds., *The historical development of physiological thought,* 243–263.

On the notion of the animal machine, see A. Vartanian, *La Mettrie's L'homme machine. A study in the origins of an idea* (Princeton, 1960). On Lavoisier and the chemical revolution, see Douglas McKie, *Antoine Lavoisier* (London, 1935); J. R. Partington, *A history of chemistry* (London, 1962), III, 362–495. Views to ca. 1840 on the supposed sites of heat production within the living body are reviewed by Everett Mendelsohn, "The controversy over the site of heat production in the body," *Proceedings of the American Philosophical Society* 105 (1961): 412–420.

Exploration of the history of metabolic studies was long the province of physiologists themselves; only recently has the subject attracted historians. While Graham Lusk's rambling "A history of metabolism," L. F. Barker, ed., *Endocrinology and metabolism* (New York, 1922), III, 1–78, will outrage the sensibilities of even the most jaded historical critic, it remains the sole sweeping survey of the subject. A seemingly narrow but in reality quite suggestive work is Charles A. Browne, *Source book of agricultural chemistry* (Waltham, 1944). A modern study of the French agricultural chemists of the 1830s and 1840s is a major desideratum; see the valuable contribution by Richard C. Aulie, "Boussingault and the nitrogen cycle," *Proceedings of the American Philosophical Society* 114 (1970): 435–479. Elmer V. McCollum, *A history of nutrition. The sequence of ideas in nutrition investigations* (Boston, 1957), covers his subject with truly the finest of sieves. Nikolaus Mani's comprehensive investigation of studies of liver function is highly pertinent to questions of general metabolism: *Der historischen Grundlagen der Leberforschung* (Basel, 1967), II, 250–369. Friedrich Müller, "Die Entwicklung der Stoffwechesellehre und die Münchener Schule," *Münchener medizinische Wochenschrift* 80 (1933): 1656–1655, is a personal account of the leading German school of metabolic research.

"Biochemistry" probably cannot properly be called a nineteenth-century science. But the chemistry of vital operations was, of course, the object of increasingly active study during the period and historical investigation of this activity also increases. The fundamental study for orientation in the subject is Frederick L. Holmes, "Elementary analysis and the origins of physiological chemistry," *Isis* 54 (1963): 50–81. One should consult also Erwin N. Hiebert, "The problem of organic analysis," *Mélanges Alexandre Koyre,* ed. I. B. Cohen and R. Taton (Paris, 1964), I, 303–325. Holmes' generous introduction to Justus Liebig, *Animal chemistry, or organic chemistry in its application to physiology and pathology,* ed. W. Gregory (reprint: New York, 1964), introduces the subject of metabolic or foodstuff "balance" studies. Broad coverage of the development of "physiological chemistry" and "biochemistry" is provided, with valuable bibliographical discussion, by Aaron Ihde, *The development of modern chemistry* (New York, 1964), 418–442, 643–670. On fermentation and chemistry, see James B. Conant, *Pasteur's study of fermentation* (Cambridge, Mass., 1952).

The perplexities regarding chemical vitalism have acquired a life of their own. John H. Brooke, "Wöhler's urea, and its vital force?—A verdict from the chemists," *Ambix* 15 (1968): 84–113, delivers the Wöhler-killed-vitalism belief a very stern knock. Berzelius' views, central to these discussions, are well treated by Bent S. Jørgensen, "Berzelius und die Lebenskraft," *Centaurus* 10 (1964): 258–281. Two essays by Timothy O. Lipman explore the ambiguities of Liebig's "vitalism": "The response to Liebig's vitalism," *Bulletin of the History of Medicine* 40 (1966): 511–524; "Vitalism and reductionism in Liebig's physiological thought," *Isis* 58 (1967): 167–185. On Liebig and vitalism, see also G. J. Goodfield, *Growth of scientific physiology,* 113–155. The chemists' approach to these matters is portrayed by Jean Jacques, "Le vitalisme et la chimie organique pendant la première moitié du XIXe siècle," *Revue d'Histoire des Sciences et de Leurs Applications* 3 (1950): 32–66 and more fully discussed and illustrated with contemporary texts by O. Theodor Benfey, *From vital force to structural formulas* (Boston, 1964). Thomas S. Hall, *Ideas of life and matter* (Chicago, 1969), chapters 25, 28–36, 43–45, provides an invaluable taxonomy of pre-twentieth century "vitalisms."

Recent scholarship has largely reduced the physicalist option to a distinctive brand of physiological reductionism. While the full legitimacy of this gesture has yet to be tested no persuasive alternative interpretations have been forthcoming; whether large-scale physiological reductionism in the nineteenth century is an historian's artefact or is indeed a central, perhaps the central, theme in the rationalization of the science is yet to be determined. The subject was brought to prominence by Owsei Temkin, "Materialism in French and German physiology in the early nineteenth century." *Bulletin of the History of Medicine* 20 (1946): 322–327 and in an outstanding essay by Paul C. Cranefield, "The organic physics of 1847 and the biophysics of today," *Journal of the History of Medicine* 12 (1957): 407–423. The reductionist interpretation has subsequently been most vigorously pursued by Everett Mendelsohn; see especially his "Physical models and physiological concepts; explanation in nineteenth-century biology," *British Journal for the History of Science* 2 (1965): 201–219. Various essays and criticisms of reductionist usage appear in the proceedings of a conference on explanation in biology, ed. Everett Mendelsohn et al., *Journal of the History of Biology* 2 (1969): complete issue No. 3. Particularly valuable in this collection are the temperate remarks of Kenneth F. Schaffner, "Theories and explanations in biology," 19–33; see also his "Antireductionism and molecular biology," *Science* 157 (1967): 644–647.

The views of the reductionists' declared enemy, the advocates of an autonomous "vital force," are concisely presented in an older anthology: Alfred Noll, ed., *Die Lebenskraft in den Schriften der Vitalisten und ihrer Gegner* (Leipzig, n.d.). Biographical material on the Berlin group is reasonably accessible. Heinz Schröer, *Carl Ludwig. Begründung der messenden Experimentalphysiologie* (Stuttgart, 1967) is an excellent account of the great teacher's research and instruction and offers as well perhaps the best available introduction to the creation during the nineteenth century of a self-consciously experimental physiology. On Du

Bois-Reymond, see Heinrich Boruttau, *Emil Du Bois-Reymond* (Vienna, 1922); on Brücke, Ernst T. Brücke, *Ernst Brücke* (Vienna, 1928); on Helmholtz as a physiologist, John McKendrick, *Hermann Ludwig Ferdinand von Helmholtz* (London, 1899). On the intellectual master of the early reductionists and one whose conceptions the latter came to deny see Wilhelm Haberling, *Johannes Müller. Das Leben des rheinischen Naturforschers* (Leipzig, 1924); Gottfried Koller, *Johannes Müller. Das Leben des Biologens, 1801–1858* (Stuttgart, 1958).

　　Discussion of positivism in physiological interpretation has properly focused on Claude Bernard; the radiation and elaboration of his views, particularly outside of France, deserve much additional investigation. The most explicit and persuasive account of Bernard's positivism is D. G. Charleton, *Positivist thought in France during the Second Empire, 1852–1870* (Oxford, 1959), 72–85. Comte's influence on French biological thought receives attention from Georges Canguilhem, "La philosophie biologique d'Auguste Comte et son influence en France au XIXe siècle," *Études d'Histoire et de Philosophie des Sciences,* 61–74. Canguilhem has devoted several essays to various aspects of Bernard's scientific research and philosophical stance; these are reprinted in ibid., 127–172. Another collection of extremely valuable essays on Bernard, his work and intellectual relationships is that of Joseph Schiller, *Claude Bernard et les problemes scientifiques de son temps* (Paris, 1967). Schiller's short article, "Claude Bernard and the cell," *The Physiologist* 4 (1961): 62–68, also merits attention. Bernard has won a great reputation for his notion of an internal physiological environment; Frederick L. Holmes relates the articulation of this conception to Bernard's interest in cell theory: "The *milieu intérieur* and the cell theory," *Bulletin of the History of Medicine* 37 (1963): 315–335, and also discusses the importance of the idea to Bernard's physiological views: "Claude Bernard and the *milieu intérieur,*" *Archives Intérnationales d'Histoire des Sciences* 16 (1963): 369–376. J. M. D. and E. H. Olmsted, *Claude Bernard and the experimental method in medicine* (reprint: New York, 1961), is a useful general assessment of the physiologist's endeavors; J.M.D. Olmsted's *François Magendie* (New York, 1944) is a superior study of Bernard's mentor and himself an exceptionally able experimental physiologist.

Index